Beyond The Edge Of Technology

50 Years Of Instrumentation and Controls at ORNL

Members of the I&C Division of Oak Ridge National Laboratory Share a Technical History Illustrating how Instrumentation and Controls was Practiced at Oak Ridge National Laboratory

Ray Adams, Dick Anderson, Don Miller, Les Oakes, and other Members of the I&C Division

iUniverse, Inc.
New York   Bloomington

Beyond The Edge Of Technology
50 Years Of Instrumentation and Controls at ORNL

FOREWORD

Fred Mynatt

I was privileged to serve as Director of the Instrumentation and Controls (I&C) Division from 1980 to 1984 succeeding Herb Hill. I have said frequently that my years as Director of I&C were the most enjoyable of my senior management assignments because a division director is responsible for the human and physical assets in a very intimate way and, of course, the I&C people and capabilities were excellent. In my previous position as Director of ORNL's Nuclear Regulatory Commission Research Programs I had a significant program in I&C for development of sensors embedded in the cladding of electric powered reactor fuel pin simulators to study the extreme conditions of a power reactor loss of coolant accident. In my earlier work in the Neutron Physics Division I also had close ties with the reactor controls work in I&C.

This report presents the history of the I&C Division and some examples of the technology. In this Foreword I will try to describe why I&C was created and I&C's role in ORNL's organization.

During World War II ORNL (then named Clinton Laboratories) had the responsibility to design, build and operate a pilot-plant-scale nuclear reactor for production of plutonium and to develop the chemical processes for extracting the plutonium from the spent fuel. These pilot plants were the stepping stones for the full-scale reactors and chemical processing plants being built at Hanford, Washington. From an I&C perspective the ORNL wartime effort required applied research and development of instrumentation and controls that did not previously exist and all the work had to be done "in-house" because the technology was new and everything was very very secret. When the first developments were successfully completed ORNL began to diversify to R&D for the production of other radioactive isotopes, the study of the biological affects of radiation, basic sciences to support development of nuclear energy and the development of different types of nuclear reactors for research, propulsion and electric power generation. Thereafter ORNL continued to develop new aspects of its original mission and to diversify into related areas of science and technology. Today ORNL is the largest (and the best!) non-weapons energy laboratory in the Department of Energy system.

The I&C Division was created as ORNL expanded and organized and it continued to grow and diversify to support ORNL's activities. When I joined I&C in 1980 the division had about 400 staff members of which about 100 were instrument mechanics and members of the Atomic Trades and Labor Council bargaining unit. The remaining 300 were engineers, physical scientists, and a small administrative staff. Since it was no longer isolated, many previous I&C staff and technologies had transferred to the private commercial sector beginning with radiation instrumentation and later to include areas such as imaging and tele-robotics. By the 1980's there were about 20 commercial companies in the Oak Ridge area that had originated in the I&C Division.

As they became available ORNL purchased commercially available instrumentation and controls and the I&C Division's role was to develop applications of the commercially available products to meet ORNL needs and to perform applied research and development of new instrumentation and controls that were prototypical, pre-commercial or customized for unique applications. About 20 percent of the I&C Division work was directly funded by the Department of Energy or the Department of Defense. The 80 percent majority consisted of joint teams for other divisions and programs of ORNL.

The I&C Division efforts in the 1980s consisted of development of advanced reactor control systems, signal processing such as noise analysis to monitor reactor and non-reactor systems, measurement and control systems for many applications, nuclear instruments for civil defense, and neutron spectrometers for neutron scattering experiments. In more recent years development of advanced "machine vision" allowed I&C to contribute to new customers such as the U.S. Bureau of Printing and Engraving.

Two areas were low-profile and rarely appreciated. The Standards Laboratory maintained secondary standards relative to National Institute of Standards and Technology and developed other standards as needed. The standards are critical to accurate measurements and control for the integrity of ORNL's research. The work of the Mechanical Design Group was rarely in the spotlight but was essential for most of the I&C developments

Since the majority of I&C work was for other divisions the question of disbanding I&C and distributing the people and resources among ORNL divisions occasionally arose. The primary motivation was the reduction of overhead cost incurred when one division purchases support from another. The advantages of continuing and improving the I&C Division were the technical excellence obtained by the mutual support of the "critical mass" of the various I&C disciplines and the flexibility of customer divisions to increase or decrease I&C expertise without adding or removing their internal staff.

Also the close relationship of the I&C technical staff with the maintenance department instrument mechanics proved valuable to both organizations.

The I&C Division/ORNL symbiotic relationship proved to work well over many years but it required I&C to keep overhead low. A key part of overhead was the "retained time" of I&C staff who did not have a full work load. In the 1980's we found that 10 percent retained time was nearly optimum. If the level was below 10% there was reduced flexibility to respond to new tasks; if the level was above 10% the increase in overhead was detrimental and I&C would need to reduce staff. This was usually done by adjusting hiring levels relative to normal attrition.

For many years the I&C Division reported to Don Trauger, ORNL Associate Director for Nuclear and Engineering Technologies and Don reported to Dr. Herman Postma, ORNL Director. Both Don and Herman encouraged and supported the I&C Division. During my tenure at I&C ORNL funded two new modular office buildings, a state-of-the-art computer design lab for design of custom integrated circuits, exploratory work in application of artificial intelligence to reactor control and a Distinguished Engineer, Dr. Robert Uhrig for work in advanced control systems. This nurturing by ORNL senior management was very important to keeping the I&C Division at the cutting-edge of measurement, analysis and control technologies.

In recent years under the new contractor, UT-Battelle, ORNL was significantly reorganized and the I&C resources were assigned to other new divisions. In a more recent reorganization key elements of I&C are reorganized into the new Measurement Science and Systems Engineering (MSSE) Division. The I&C Division has a proud history and, under the new organization, a bright future.

Following his assignment to I&C Dr. Mynatt was appointed to: ORNL Associate Director for Nuclear Engineering and Technology in 1984; ORNL Associate Director for Reactor Systems in 1987; Martin Marietta Energy Systems Vice President for Compliance, Evaluation and Policy in 1990; Lockheed Martin Energy Systems Senior Vice President in 1994; Lockheed Martin Energy Systems Vice President for Supporting Operations and Compliance in 1996 until retirement in 1997.

Contents

Acknowledgments

This history of ORNL Instrumentation and Controls work could not have been written without the help of a great number of people. Of course, there are people who have helped over the years, whose names I have forgotten, and for those omissions I am truly sorry. I am grateful to Jim Hardy and his secretary, Judy Hill, who sponsored my residence (so I could search for the facts) in Bldg 3500 (My office was in that building for many of the years I spent in I&C). Also, I am grateful to Kathy McIntyre, (formerly the secretary to I&C's last Division Director, Dan McDonald). Before her retirement, Kathy was secretary to Ted Fox, the Division Director for the Engineering Science and Technology Division, which inherited many of the I&C professionals, when the division was eliminated by substantial reorganization, about a year after UT-Battelle obtained the operating contract.

Among those who helped with the gathering of material, have been Phil King III and especially, Dianne Griffith, formerly of ORNL Lab Records. Dianne did many more searches than I was directly aware of, and who encouraged Missy Pointer, to dig through the IRC (Inactive Records Center) in ORNL storage, for some of the most interesting documents that pre-date the formation of the I&C Division. More recently Missy helped by finding some elusive documents stored in the deep recesses of the ORNL and (DOE organization's) DTIE files. Most of the photographs and many of the drawings in this volume were unearthed by Joy Anderson, in the Communications and Community Outreach organization of ORNL, Joy seems to have a knack for remembering anything she has ever seen before, and she truly lived up to her name as a tireless "Joy" to work with.

The authors of the material herein are to be commended for their hard work. Also, this book is overdue in terms of underrepresenting or omitting reports of the fine work done by several persons who are now deceased. The authors regret not being able to tell the contributions of those former members of the I&C Division. All of the ORNL and Y-12 photos that appear in this book are in the "public domain" as that is the nature of such institutions as ORNL and Y-12, funded as they are by "public" (i.e. US Government Contracts) money. All other photos and illustrations were created by the authors, and/or the editors.

The agreement of Barbara (Bobby) Lyon, to edit the volume was especially fortuitous. The goal of this document is to communicate clearly and consistently the significance of the achievements of ORNL's Instrumentation and Controls Division researchers over the first 50 years of the Laboratory's distinguished history. Thanks to excellent editing by Barbara Kennedy Lyon, our document, penned by a diverse group of retired researchers, has (hopefully) found its voice for transmitting our messages on the wavelengths of our readers. We acknowledge Bobby's gifts for finely tuning our words so that our intended messages are understood. Bobby Lyon is founding editor of the Oak Ridge National Laboratory REVIEW, ORNL's prestigious research and development magazine, see (www.ornl.gov/ORNLReview/). From 1967 through 1982 she edited the magazine founded by then ORNL Director Alvin Weinberg. Today she is a well-respected free-lance editor who writes and edits with grace and style. Bobby lives in Oak Ridge and enjoys visits with her two talented children, grandchildren and great-grandchildren.

Bobby capably managed the content of the *ORNL REVIEW* from its founding in 1967, until she retired in 1982 - passing along the editorship to Carolyn Krause, whom she trained and who was the editor until December 1 of 2008. Carolyn herself was very helpful in the gathering of information for this book. She provided much of the copy for several of the chapters, which contain technical papers by I&C Division authors. These papers have not been published until now. This is partly due to a loss of funds for publishing them as a special Instrumentation issue of the *ORNL REVIEW*. They are among the articles forwarded to me by Carolyn Krause for inclusion in this book.

Ray Adams - Oak Ridge, TN, January 2009

Dedication

Integrity:

Any close look at the history and operations of an organization has got to come to grips with the basic integrity of the organization. The integrity of ORNL, it has always been the observation of the principal author -Ray Adams, has always been of the highest order. Furthermore, it has been the privilege of the authors of this work, all of who worked in the I&C Division, to have observed that the management of the Division practiced the highest integrity in dealing with their clients and employees. One of the authors, who was a member of several car-pools from home to work, over many years, had difficulty believing the tales, told by other car-pool members, of "broken management integrity." "Nothing like that ever happens in my Division," was his response as these tales unfolded. Integrity begets integrity and it is believed that while it existed, the I&C Division contributed not a little, to the basic integrity of ORNL.

This volume of *"BEYOND THE EDGE OF TECHNOLOGY - 50 YEARS OF INSTRUMENTATION AND CONTROLS AT ORNL,"* is dedicated to the following persons- all gentlemen of highest integrity:

Don Trauger- ORNL Associate Director during most of the time of the existence of the I&C Division.

I&C Division Directors
- C. J. Borkowski   - F. R. Mynatt
- B. G. Eads   - D. W. McDonald

Introduction - by Ray Adams

What IS Measurement and Control?

In 1883, William Thompson[1], the eminent Scottish Physicist, before he was knighted in 1892, as Lord Kelvin, declared that:

> **"In physical science the first essential step in the direction of learning any subject is to find principles of numerical reckoning and practicable methods for measuring some quality connected with it. I often say that when you can measure what you are speaking about, and express it in numbers, you know something about it; but when you cannot measure it, when you cannot express it in numbers, your knowledge is of a meager and unsatisfactory kind; it may be the beginning of knowledge, but you have scarcely in your thoughts advanced to the state of Science, whatever the matter may be."**

Lord Kelvin made his own contributions to numbers, among which were: That which we know as the freezing point of water, thirty-two degrees Fahrenheit or zero degrees Celsius, is 273.15 on the Kelvin scale, where zero is reserved for *absolute zero,* the point at which a system has minimum possible energy. Lord Kelvin's declaration about the principles of measurement were a commonly stated modus operandi of members of the I&C Division.

So much for measurement, which is done with *measuring* instrumentation. The science of control is related to holding constant, or programming a variation of one or more of the important parameters of a scientific or engineering study. When I joined the I&C Division, shortly after it was formed in 1953, I thought that as a member of the Instrument Department, that the other Department in I&C, namely the "Controls" Department would do all of the controls work. However, it soon became clear to me that what they were specializing in was *Reactor Controls*, and that there would be plenty of other types of control for me to engage in. I was a member of the Applications Section, which would shortly be named the Process Control Section, doing measurement and control of physical processes – primarily Temperature, Flow, Pressure, and other physical variables.

One mark of a professional organization is whether or not there is a national or international professional society that its members belong to. In the case of the I&C Division, over the years, its members found professional recognition in at least three national Professional Societies, The ISA, the IEEE, and the ANS.

The Instrumentation Systems and Automation Society

The ISA[2] started out shortly after WWII, in 1945, as the Instrument Society of America. In 2000, its Board of Directors changed the name to Instrumentation, Systems, and Automation Society, in recognition of its expanded role, worldwide. Oak Ridgers have always played a large role in the ISA. Indeed, in 1954, a former member of the I&C Division, Warren Brand, was elected to be National President of the ISA. Over the years also, there have been a goodly number of National Fellows, elected from Oak Ridge, to that distinguished membership grade in the ISA. Another distinguishing feature of the ISA, has been its inclusion of the Instrument Mechanics and Technicians – at ORNL, those employees were classified in the bargaining unit (Labor Union). In the ISA, as in the I&C Division, the partnership of the Engineers and Scientists, with the Instrument Mechanics and Technicians has always been strong and contributed greatly to the services provided the research staff of the Laboratory.

The Institute of Electrical and Electronics Engineers

The IEEE[3] (Institute of Electrical and Electronic Engineers) evolved from a merger of the IRE (Institute of Radio Engineers) and the AIEE (American Institute of Electrical Engineers) in1961. Prior to this merger, I&C members were divided in their affiliation with either or both groups. The IEEE has a number of specialty "Societies" in which various I&C members published papers. Over the years, I&C members submitted and published papers in, among others, the Circuits and Systems Society, the Control Systems Society, the Computer Society, and the Nuclear and Plasma Sciences Society, among others.

The American Nuclear Society

For the ANS (American Nuclear Society), December 11, 1954, marks the ANS's historic beginning at the National Academy of Sciences in Washington, D.C.. The National Headquarters office of the ANS moved around some during its early years. ANS's first "home" was in space provided by the Oak Ridge Institute of Nuclear Studies in Oak Ridge, Tennessee. Today, it continues to be an organization of scientists, engineers, and other professionals devoted to the peaceful applications of nuclear science and technology. Its 11,000 members (in 46 countries) come from diverse technical disciplines ranging from physics and nuclear safety to operations and power, and from across the full spectrum of the national and international enterprise, including government, academia, research laboratories, and private industry. One of the early Presidents of the ANS was Alvin Weinberg, an early Director of ORNL. He served ANS as president in the 1959-1960 term.

The Early Days of ORNL

In the early days of ORNL (in 1943, when it was called Clinton Engineering Labs), and its operation by DuPont and Monsanto, it existed for the purpose of studying the chemistry of Uranium and for purifying the element plutonium, a product of the Graphite Nuclear Reactor (X-10 pile). Of course the study of the chemistry of Plutonium was done as well.

The Early Days of I&C

When the I&C Division was formed in February of 1953, it consisted of an Instrument Department[4] and a Controls Department. The Instrument Department had evolved in the Plant and Equipment Division, and earlier from the engineering groups that were formed to support the chemical re-processing of the irradiated fuel in the Graphite Reactor. Somewhat later within ORNL, there were instrument engineering and research groups set up to support researchers - notably in the Chemistry Division, from which came the first Director of the I&C Division, C. J. Borkowski. These groups became a small group of development specialists and/or were merged into the Instrument Department of the I&C Division.

The Controls Department was newly formed, from physicists and engineers who had been in various ORNL organizations, engaged in Reactor Controls. The control and safety aspects of nuclear reactors had to be evolved from basic principles of the relatively newly discovered chain-reacting physics and did not have the long established basis of control of chemical and physical plants that had attended the application of other types of measuring systems. Hence the philosophies of nuclear reactor safety and control systems are relatively new. So important was this aspect considered, that the I&C Division assigned an engineer full time to the ORNL Nuclear Safety Information organization when it was formed.

What Were the Benefits to ORNL of an I & C Organization?

Many, even among some of the ORNL management, were unaware that their research members greatly benefited from the existence of the I&C Division. Yet ORNL management, at the Associate Director level was keenly aware that this organization performed many valuable measurement and controls services. Whenever funds got tight and the ORNL organization sought to reduce waste and cut out overhead charges, the I&C Division would have to justify its existence by assessing how every bit of its existence was derived. In the early days, I recall having to look closely at every job in which I, or my group was supporting the research effort and document that support from the research Divisions, in minute detail.

Later on, as research funds were allocated more directly to I&C research, this exercise became a bit easier. However, Don Trauger, one of the most respected ORNL Associate Directors (under whom the I&C Division was assigned) once told me that he had to defend the I&C Division "many times" against moves to weaken its organization. This often occurred when there would be a move to dissolve it, or to cut off a portion of its structure. The I&C Bargaining Unit employees (Mechanics and Technicians) were frequently the object of such threats. The other bargaining unit employees (e.g. in the Plant and Equipment Division) perceived that a special status was held by the I&C bargaining unit employees, by virtue of their working closely with the Engineers and Scientists in the I&C Division.

Indeed, the Engineers and Scientists in the I&C Division took pride in and felt the support of ORNL's research mission as enhanced by this close relationship with the I&C Mechanics and Technicians. They were the "doing right hand" of the I&C Engineers and Scientists and made numerous contributions to the success of I&C efforts on many projects. They were often the "front line" of support in keeping jobs going, after an installation was designed and installed and for maintenance along the way. They often referred to their work as "from the cradle to the grave."

How This Book Was Written

Ray Adams

In the late 80's it was decided by upper management that each of the Research Divisions of ORNL should write a History of their organization to be published by the 50th anniversary of the Laboratory – 1993. The I&C Division was invited, as were Analytical Chemistry, Chem. Tech, Metals & Ceramics, Chemistry, and others. Those efforts may be seen at the ORNL Web Pages under "History". However, the I&C Division's effort never got beyond draft stage. I have read the draft and it is of very limited usefulness. I am glad that it was never published.

After I retired, I decided that there ought to be a history written of the I&C Division. There were so many technical innovations by so many people, and ORNL had benefited to such a great extent during my tenure that I thought it somehow ought to be publicized, especially for other technical people. Many have contributed to this volume. I was in the I&C Division almost as soon as it was founded, staying in the division for over 32 years. I was acquainted with most of the members of the division up until I took "early" retirement to teach in the Electrical and Computing Engineering Department of the College of Engineering at the University of Tennessee, in Knoxville. So, Having had close ties to I&C Division people for many years, I decided that some of my fellow retirees could help write the history of the I&C Division. This book is the result.

The Organization – Over The Years

I&C Division Directors and their tenures are as follows:

E. D. Shipley, an Associate Director of ORNL, served as acting Director from the founding of the I&C Division, in February of 1953, until C. J. Borkowski agreed to be the I&C Director, in February of 1954. Ed Shipley was an avid gardener and in that first year of the I&C Division's existence, he brought his lunch to eat in the (New Bldg 3500) lunch room. He brought two containers of Strawberries and Cream each day.

**Picture
Not
Available**

ORNL News3446

Casimer J. Borkowski agreed to be the first I&C Division Director and served from February 1954, until his retirement in 1977. He maintained his interest in advanced instrument development and was a leader of a small group of development specialists throughout his tenure as Division Director.

H. N. Hill was appointed I&C Division Director in June of 1977 after an exhausting search (Throughout the Union Carbide Corporate Organization) for suitable candidates. He served as I&C Director until April 1, 1981. That period saw the retirements of several long-time emplayees and the consequent reorganization of several of the I&C Departments.

ORNLHerbHill

Fred Mynatt was appointed I&C Division Director to serve from April 1, 1981. He was promoted to Associate Lab Director for Nuclear Engineering and Technology in 1984; ORNL Associate Director for Reactor Systems in 1987.

ORNL2319-88

Bill Eads was appointed I&C Director to serve from April 1, 1984. In 1992 he accepted an appointment from the Governor of Tennessee to assist with Science for the State of Tennessee. He went to the state capitol (Nashville) and was active in directing Tennessee Science initiatives until he retired. He still lives in Nashville

ORNL 394-93

ORNL 1050-98

Dan McDonald served as Acting Director of the I&C Division from Jan 1, 1993 and became Director effective Jan 1, 1994. He served until the I&C Division was eliminated through re-organization, by the New Operating Contractor on October 1, 2001.

In March of 2008, portions of the previous I&C organization were formed into a new Division, called the Measurement Science and Systems Engineering (MSSE) Division. This re-incarnation of (portions of) the old I&C Division is headed by Ken W. Tobin. His tenure began March 1, 2008.

ORNL 07599-2001

Original Department Heads

In the early days of the I&C Division, from its founding in 1953, until about 1980, there were but two Departments, The Instrument Department (carried over from the P&E Division), and a new Reactor Controls Department.

Charles S. Harrell was head of The Instrument Department, and is shown at right.

ORNLC.S.Harrill

The new Reactor Controls Department was headed by E. P. Epler. He is shown in the picture at the left

ORNLNews5541

15

About the Authors of this book

There are more than fifteen contributors to this book. Each is a present or former member of the Oak Ridge National Laboratory (ORNL) and the former Instrumentation and Controls Division of ORNL Some of the contributors submitted only a few words. Some only contributed a few sentences in oral conversations. The principal contributors are:

Ray Adams -
Principal Author/Editor

Ray is the organizer of the group of contributors and is principal editor/author. He was a key member of the Instrumentation and Controls Division of ORNL, having served as a keen observer of that organization, through his 32 + years of service as an engineer and manager. In addition to authoring numerous refereed technical papers and in-house reports over the years, he is the author of the chapter (87 pages) on Instrumentation in the Nuclear Industry, in the book: "Instrumentation in the Processing Industries" edited by Bela G. Liptak, published by Chilton: -1973: ISBN 0-8019-5659-5

Education: B. S. Eng. Physics, University of Colorado (1951)
M. S. Eng. Science, University of TN (1961)

Professional Experience:

Instrument Engineer, Oak Ridge Gaseous Diffusion Plant, (1951-53). At ORNL, (1954-86), he was a Development Engineer and Group Leader in the Instrumentation and Controls Division, at one time supervising about 50 Development Engineers engaged in Digital Computer, Instrument Development and Metrology. He left this career in 1986, to teach Sophomore Electrical and Computer Engineering students in the College of Engineering, at the main campus of the University of Tennessee, Knoxville, TN (1968-94).

He was a consultant on Intellectual Property in his own firm, from 1983-97.

Professional Affiliations:

Institute of Electrical and Electronic Engineers (IEEE); Senior Member;

Instrumentation Systems and Automation Society (ISA), Fellow.

Numerous (refereed) technical publications and two patents

Dick Anderson - Author/Editor

Dick was hired (by Ray Adams) to head the Metrology group and he progressed to become the Chief Scientist of the I&C Division.

Education: B. S. Chemistry, MIT (1958)
PhD. Physical Chemistry, Rice University (1964)

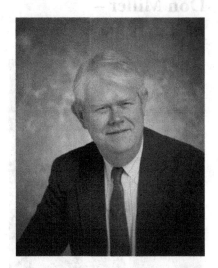

ORNL03600-93b

Professional Experience:

Worked in the Temperature Section of the Heat Division at the National Bureau of Standards from 1963 to 1971 on the NBS High Temperature Gas Thermometer project as well as on high temperature platinum resistance thermometers. From 1971 to 1974 worked in the Heat Division at the Physikalisch-Technische Bundesanstalt (West German Bureau of Standards) on designing and building a low temperature gas thermometer. From 1974 to 1989 worked in the Instrumentation & Controls Division at the Oak Ridge National Lab as head of the Metrology R&D Laboratory and Group Leader for Measurements & Sensors.

Professional Affiliations:

From 1989 to 1992 worked in the Metrology Dept. and Sensors and Process Integration Divisions at SEMATECH in Austin, TX. In 1992, returned to ORNL as Chief Scientist of the Instrumentation and Controls Division.1998-1999, co-ordinated and ran road map sessions for the technical divisions of ISA, American Physical Society: Fellow & Former Chair of the Topical Group on Instruments and Measurement Science (1992) Instrumentation, Systems and Automation Society: Fellow & Chairman of the Metrology Committee of the Test and Measurement Division, Co-coordinator of ISA National Roadmap;

Don Miller –
Author/Editor

Don was an instrument techni-cian (hourly pay rank) in the I&C Division, worked with Ray Adams and was promoted through the ranks to the salaried position of De-partment Head of the Maintenance Management Department of the I&C Division.

ORNL02587-90e

Don is the primary contributor and author of chapters in this book, relating to hourly worker and in-house ORNL support work. He holds a US patent for "a device and method of measuring the coefficient of performance of residential heating and ventilating systems." He was Department Head of the I&C Maintenance Management Department from 1984 to 1994.

Previously he was, in succession, an hourly Instrument Technician, Engineering Technologist, Staff Engineer, Group Leader and General Foreman in the I&C division.

Professional Experience:

He chaired a five-plant committee for Martin Marietta for three years focused on innovation and empowerment of facilities staff to share technology and procedures using predictive techniques for machine reliability.

He provided encouragement and focus to a team of maintenance staff who ultimately developed maintenance software, called the MAJIC system.

This software has been judged "best of the ORO contractors" by the government oversight officials. He retired in 1994 to take a position with SAIC as a Senior Research analyst providing site and facilities planning services to ORNL and Y-12. He has authored a number of technical papers and has made several presentations of Measurement and Controls technology and effective maintenance organization for such.

Lester Oakes - Author/Editor

Les worked as a development engineer in the Reactor Controls Department of the I&C Division. He became a Group Leader, ORNL Corporate Fellow, and the head of the Reactor Controls Department of the I&C Division.

ORNL04066-84

Education: B. S. EE., University of TN (1951)
M. S. EE., University of TN (1962)

Professional Experience

Control Engineer, Fairchild Engine and Aircraft Co. 1949-1951.

At ORNL he began in 1951, as a development engineer assisting in the development of a computer for use in the dynamic analysis of reactor control and safety systems. Later he became a group leader responsible for the design of controls and systems for reactors built at ORNL. He then became head of the Reactor Controls Department and Associate Director of the Instrumentation and Controls Division. He was an ORNL Corporate Fellow. He was a consultant for both DOE and NRC and was a contributor to the DOE analysis of the Chernobyl Reactor (and its accident). He spent a year on loan to the Electric Power Research Institute assisting in the analysis of the Three Mile Island Reactor accident. He retired from ORNL in 1991.

Professional Affiliations:
Life Fellow of the Institute of Electrical and Electronic Engineers.

Numerous technical publications on reactor control and protection. Two patents on reactor control devices, including an automatic startup system.

Introduction To Section 1

By Ray Adams

This section contains a summary of the work on Measurements and controls that predates the formation of the Instrumentation and Controls Division, per se. To some extent, the work reported herein is inferential, but it is nevertheless, based upon the reports, organization charts and personal interviews with persons who were in Oak Ridge during those early days.

Foremost among those individuals was Casimir Borkowski, the first division director of the Instrumentation and Controls Division. Those of us who worked in the I&C Division caught many glimpses of the man and the work of the Early Manhattan Project days, as we were privileged to discuss them with a person who had been with the project in its earliest days. Being in the earliest cadre of scientists who created the first atomic pile at the Chicago stadium, Cas was a founding member of the group that created the instruments that went beyond the edge of technology. These included anything that measured and controlled the newly found source of energy Einstein referred to in his famous equation $E=mc^2$

The second person I sought for insight into those early days, was Ellison Taylor. Ellison was first of all, a Chemist. Secondly, he was an administrator who guided the work of chemical research at ORNL from the vantage point of having been at ORNL in its earliest incarnation as Clinton Labs and who was instrumental in getting Cas Borkowski to be the I&C Division's first division director. Ellison was not only a fine chemist, but he maintained his interests and skills well after retirement. I recall teaching a (beyond retirement) class in how to use computers and how to access the Internet, which he attended – showing other students in the class that he was not too far beyond retirement to learn several new things.

There was no "Nuclear Technology" when the Manhattan Project was started as a war-time effort to try to make a nuclear weapon before the enemy could. The "edge of technology" at that time consisted of mechanisms that were essentially those to be found in college science halls. Because the very foundation of ORNL (initially called Clinton Laboratories) was as a pilot plant for the Hanford (production) Reactors and their attendant chemical separation and extraction of Plutonium from the highly radioactive reactor fuel, much of the instrumentation was "beyond the edge of technology". Many of the instrumental measurements had never been done before – certainly not in production-scale, for either process control or for laboratory purposes. And much of the work was "Secret". I had occasion to telephone one of the early Oak Ridge persons, who showed up on an organization chart at Oak Ridge. I had known him, through a technical society, as an engineer with one of the instrument companies, associated with process control. When I started to ask him about his involvement.in Oak Ridge, he said "That stuff is all classified secret." I replied that it had long since been de-classified. He replied, "You say it has been de-classified, but as far as I am concerned it is still secret." Our conversation is over – and it was. I later sent him a document pertaining to those de-classification efforts, but never talked to him again – our conversation was truly "over". Such was the secrecy of those early days, when we were attempting to win the war.

Chapter 1

The Nature of Early I&C Work at ORNL

In The beginning of Clinton Laboratories

When the Manhattan Engineering District created the Clinton Laboratories, in 1943, it was for the purpose of its being a pilot plant for the Hanford Engineering Works in Hanford Washington. That facility was to produce the fissionable element Plutonium in sufficient quantity to produce enough Plutonium for Bombs to be made. Early experiments and many calculations had shown that although Uranium 235 - the product of the huge Uranium separations plant in Oak Ridge, at K-25 and the electromagnetic enrichment plant at Y-12, could make a bomb from the fissionable U isotope 235, plutonium was a better candidate. Of course, the chemistry of the new element Plutonium also needed to be studied, so that the X-10 Graphite reactor and the chemical separation pilot plant's first product went toward the study of the chemistry of Plutonium.

So, as a pilot facility, Clinton Laboratories first task was to create Plutonium in the Graphite Reactor at facility X-10. Upon demonstration of a successful Plutonium process, a chemical separations plant would be needed. The chemical separation plant was designed so that all separation chemistry would be isolated from the operator areas because radiation levels expected in the chemical plant were to be much higher than human operators could safely work. Therefore that chemical separation of Plutonium from the mix of Uranium 235 and 238 and radioactive fission products had to be done remotely, in "hot cells." This radiochemical process was to be a pilot plant for the radiochemical separation at Hanford, WA. There is a well-written report of the history of the ORNL chemical separation plant[5].

The Chemical Pilot Plant

When the X-10 plant was laid out and the buildings were assigned numbers by the construction contractor DuPont, the Reactor building was given number 105 and the chemical (pilot) plant building next door, was given number 205 a building that was later re-numbered building 3019. Little known to many of ORNL's present scientific staff, there was a connection between the two buildings. At the rear of the reactor, there was a chute by which recently irradiated fuel slugs, pushed out of the reactor. These fuel slugs had achieved their purpose of containing the Uranium fuel so that some of it was "elevated" in the reactor, to the new element Plutonium, could then be taken to their next processing step. Those fuel slugs fell into a container under water in a water-filled canal that led over to the pilot plant next door - Bldg. 205. In this manner, the product of the reactor was conveyed through the water, which provided shielding, to become the feed for the radio-chemical processing (pilot) plant. The building arrangement is shown in the photo below.

FromORNLHist152

View of the construction of the Graphite Pile Building and the Chemical Pilot Plant. A canal between the two buildings allowed the transfer of irradiated fuel slugs from the reactor to the chemical processing plant (under water for shielding).

The Product of this plant was initially Plutonium and the know-how for building the Hanford extraction plants. Later on, this facility (always referred to as the "Pilot Plant") was used for several types of chemical extractions, Engineered by the Chemical Technology Division of ORNL, as ORNL continued to chemically process highly radioactive materials. The original instrument panels are shown here.

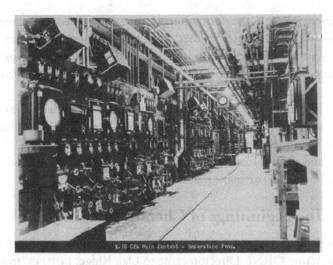

X-10 CEs Main Control - Separation Proc.

ORNLHist185

These Instrument Panels give evidence of a "process control" type of instrument applications work, shown in an early Clinton Laboratories photo.

An early organization chart (the earliest one found, dated 2/25/44) shows an Assistant Superintendent over Instruments, in the person of W. P. Overbeck and 53 others. That was a significant number of people, as the whole of Clinton Laboratories at that time was a total of 1099 employees.

Because the nature of the chemical extraction process was highly radioactive, all of the process measurements had to be made remotely and their signals had to be transmitted to the panel board via safe-remote means. In those days the means was pneumatic transmission that had been worked out in the U.S. chemical industry. This requirement necessitated extra instrumentation "transmitters" and an extra remote "transmitter panel" located in a gallery above the process vessels yet shielded from the "hottest" parts of the process. This remote transmitter rack could be attended by instrument mechanics if they made their visits brief, so as not to be exposed overly long to the radiation in this part of the building. The instrument engineers and technicians became adept at devising means to measure liquid levels, pressures and temperatures through these remote sensing instruments that then transmitted the measurements to the operators panel board indicators and recorders.

The demands of the chemical extraction and its requirement for instrument technicians and engineers doubtless contributed to a conflict with the "research" nature of the laboratory. It wasn't long before there began to be cries that the research chemists and physicists could not get "instrument service." This complaint was brought before the research council meeting and led eventually to the formation of one centralized Instrumentation and Controls Division. Prior to that, however, the instrumentation services at the laboratory were placed under a single department -The Instrument Department of the Engineering Maintenance and Construction Division.

Thus, the earliest Processing plant in Clinton Laboratories was a chemical plant for the separation of Plutonium from the Uranium and (radioactively 'hot') fission products, in the Pilot Plant, building 205. This plant had measuring and control instruments that needed maintenance and adjustment by an Instrument Department that had 54 persons in it.

The beginnings of Chemical Engineering

Floyd Culler, director of the Chemical Technology Division, and eventually Acting ORNL Director, came to Oak Ridge, in 1943, from Johns Hopkins University. After a stint at the Y-12 (Electromagnetic Separations) plant, he started at ORNL (formerly Clinton Laboratories) in 1947, as a design engineer for nuclear fuel-recycling plants. He advanced quickly to become the head of the Chemical Technology Division. Under Culler's direction, ORNL became the AEC's Chemical Processing design center.

There was a close relationship between the Chemical Technology Division and the engineers in Instrumentation and Controls work as the (radioactive) chemical processes required extensive application of measurement and controls technology for the remote sensing and controls required by those processes.

Reactor Controls

In other aspects of the instrumentation and Controls work at Clinton Laboratories, the control panel for the Graphite reactor was as shown in the figure below.

ORNLHist184

The reactor operator controlled the reactor by keeping the light beam of a galvanometer centered in a window (directly under the clock to the left in this picture). He did this by manipulating control rods that moved in response to a control handle that acted to insert or withdraw control rods in the reactor.

Chapter 2

Clinton Laboratories Instrument Maintenance (1943 to 1953)

Don Miller

Instrument work at Clinton Labs (It later became ORNL) was a vital part of the new Manhattan Engineering District isotope separation activities. The previous chapter showed pictures of the panel boards of several of the early installations (Chemical Plants and Nuclear Reactors) and the installation. The maintenance of those instruments was the domain of the I&C Maintenance technicians. Design, fabrication and use of first generation reactor controls, chemical plant controls as well as fission measuring and recording equipment and health related monitoring equipment required the work of skilled craftsmen from the beginning. The Clinton Engineering Works plant was laid out with production and training reactors, and hot cells located east of (what is now) Third Street and north of Central Avenue during 1943. This area was immediately set up as an exclusion area requiring a (Q) clearance[6]

The buildings east of what is now called Third Street were surrounded by a fence which extended from behind 706-A north to enclose building 101 and the water treatment plant. The fence then extended west to the 706-B, metallurgy annex, and then south to connect with the Q Area guard gate on Central Avenue. There were four guard towers; two south of the Laboratory, one east of the Laboratory and one north of the Laboratory, all inside the outer type one fence. There were two field canteens, one just south of the Pile Building and one just north of the Carpenter Shop. A Type One fence surrounded the entire plant from Bethel Valley Road to White Oak Creek. A three-foot wide path was inside the outer fence for the Patrol footpath. The map that shows all of this was classified Secret, but that classification was canceled in 1952.

The aviation navigation map shown below, depicts a restricted area in Anderson and Roane counties, which was not approved for aircraft overflights.

There was one incident in the early '50's, in which a small plane wandered into this area,

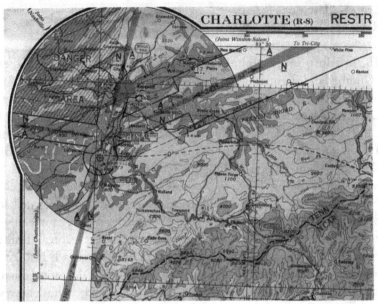

Early (1944) aviation map that shows the exclusion area prohibiting aircraft overflights, over the Oak Ridge and Manhattan Project areas.

causing Air National Guard units in nearby Knoxville, to "scramble" to intercept the threat. The fighter planes required the small plane to quickly land (on a plot being readied for a parking lot of a local mall in the city of Oak Ridge) where the pilot was met by the 'government' security patrol on the ground.

The following reports of activities are from recall of personal conversations plus reference to the organization charts of those days. In the 1943-53 time frame W. J. Ladniak was the supervising Engineer of the Service Branch. P. D. Schnelle was the staff Engineer leading the support work. They initially serviced the newly completed Graphite Pile; building 105 and Pilot Plant; building 205 plus all 100 and 200 buildings. These building numbers were selected by DuPont according to function. Later Union Carbide changed the building numbers to correspond to geographic location. The instrument mechanics assigned in 1947 and listed on Monsanto Chemical Company seniority list were:

Delbert G. Davis (3/44 hire date)
John B. Frisbie (4/44)
Robert R. Hall (from Physics ~ 10/11/43)
Andrew J. Isenberg (4/44)

29

James C. Maxwell (3/44)
Willoughby Ragan, Jr. (12/14/43)
Leon H. Timmons (8/44)
Raymond W. Tucker (11/44).

The first Instrument Technician hired was Robert Russell. He had come from Cumberland, Kentucky to the Physics Division. He was about 35 years old when hired and after a short assignment in Physics was hired by the Instrument Department. He retired in 1974, making him about 65 at that time. In February of 1946 Ray Tucker came to work from TVA (accounting dept) upon a recommendation from his supervisor at TVA who had a friend in the employment section of Dupont Chemical Company. He had been discharged from the Air Force in November of 1945 as a Radar Mechanic (862) and had a "Q" clearance and as he understood it they needed some warm bodies. My supervisor knew that I would leave TVA for college in September so he said why don't you go across the street at break time and talk to Mr. J. A. Cook. He will be interested in you. The rest is history. The reason he is listed as having a 1944 hire date is the credit the union obtained for veterans after Carbide took charge. Ray Tucker said, "The fellow who hired me was a Dupont manager of the instrument section at that time but I have forgotten his name. I don't think we were called instrument people at that time. The 2506 building was called 'the castle'."

Charlie, (Little Tuck, Ray's brother) hired in the stores department after he was discharged from the Navy. I don't know the date. He bid in as an apprentice in the department. Both Ray and Charlie Tucker were Knoxville born people and attended local schools.

Instrument maintenance activities were set up in building 717-B (later to be designated 2506) in the L clearance area west of Third Street. This building housed instrument maintenance supervisory personnel, a fabrication shop and a maintenance group. Also located in this building were a small dedicated machine shop, a stores facility for instrument parts, the Tube Shop, Drafting and an instrument development group reporting to engineering supervision in the Engineering and Maintenance Department. In house instrument design was driven by the absence of suitable equipment from the marketplace. The Tube Shop designed and fabricated radiation detectors such as Geiger-Mueller detectors. Several field shops were established as time went on. Shops were located near the Chemistry Building, Graphite Reactor Building, the Pilot Plant, and in the Y-12 area occupied by the ORNL Biology research effort. In 1949 Cardwell was appointed head of the E&M Division and Charles S. Harrill became the Instrument Department Superintendent. Ray Tucker remembers working for Ken Forrestal, who is listed on a 1950 organization chart responsible for instrument construction.

Taken in 1944, this picture shows the Instrument Department as it was then. Both technicians and engineers are shown. There are 50 people shown, and it is likely the group is as inferred on the 2/25/44 (earliest) Clinton Laboratories Organization chart as W. P. Overbeck (supervisor) and 53 others.

George Holt was the development engineer in that group. In 1947 Bill Bird joined this shop. He remembers having banks of vacuum tube aging racks, a facility known as the "Tube Shop". The tubes were to be used in Geiger Counters, proportional counters, BF-3 counters and later the ORACLE computer tube racks. Bill also remembers D. D. Walker teaching in the Apprentice Program, and George Holt being very helpful to the shop. Bill noted that Monsanto wanted to give all hourly a 25-cent per hour raise, but Carbide and Carbon Chemical Company (C&CCC) was awarded the contract before that occurred. From an hourly-worker point of view, the decision to shift from Monsanto to C&CCC was to herald significant policy change. The most significant would be the invitation of organized labor to come into the federal facility. Also joining the shop late in 1947 was Gerald Hamby. Gerald was one of the initial Instrument apprentice class members. He remembers that it was well organized and planned for the work to be done. Gerald remembers the helpfulness of John Horton when difficult technical issues arose. Herb Linginfelter was hired in 1948 and worked for Bob Toucey initially. Herb remembers Myron Fair teaching math and radiation terms. Another west area shop was headed by Al Novak with P. S. Buckley staff Engineer. This was for the 700 and 800 buildings. Other field maintenance shops were set up near the Chemistry Building, the Graphite Reactor Building, the Pilot Plant, and in the area of Y-12 occupied by the ORNL Biology staff. All instrument work was initially under the mechanical Department of the Engineering, Maintenance and Construction Division (later to be designated Plant and Equipment Division). This arrangement was changed with the organization of the Instrumentation and Controls Division in 1953.

Instrument Mechanics assigned in 1947 were: Ocee C. Cole (12/44), J. D Boone (8/44), J. D. Eddlemon (11/44), Paul W. Hill (11/44), Robert T. Pratt (12/44),Alton D. Williams (10/13/43), and P. P. Williams (11/44).

Later the Circuit Shop was under the leadership of R. E. Toucey, and had grown to eleven hourly on staff: Engineer D. C. Thompson, Foreman - K. N. Forrestal; Instrument Mechanic - Ancil E. Ball (6/45), John M. Groover, Jr. (5/49) Samuel U. Hamric (5/46), Herman J. Hurst (6/45), Earl C. Moore (11/44), John A.Oliver (12/46), M. M. Oppegaard (12/44), Earl W. Sparks (11/44), Don N. Todd (10/46), Charles D. Vail (3/45), and D. D. Walker (12/44).

All of the craft-like instrument work performed during the war years was accomplished by engineers and scientists, and nine Instrument Mechanics (later to be designated Instrument Technicians). Prior to 1940, pneumatic tubing and transducers transmitted 95% of all instrument signals and controls. Thus the term mechanic was a natural call. As the technology of measurement and control of industrial systems matured, the term technician emerged as applicable. A parallel support group was the storeroom attendant group. We don't know if any were assigned to instrument material delivery, however nine were listed on the seniority list by war's end. A later seniority list noted two Instrument Checkers. Beatrice Burgner was hired in 1944 and Benjamin Ford was hired in 1967. Major construction projects included the Graphite Reactor, completed in 1944, the Oak Ridge Research Reactor built in the 1950's. Each of these projects had a significant contribution from the Instrument Department. Instrument Technicians were challenged throughout this period to keep abreast of technology, specialized procedures being developed both by I&C engineers and other divisions in their development work. Along with these technological and procedural changes came new fabrication techniques, some of which will be noted in later chapters. Although in a relatively primitive state of development, the health physics and industrial hygiene activities strongly impacted field work for I&C maintenance.

Also of historic interest was the free flow of personnel between plants. A September 1968 memo from C. S. Harrill announced the transfer of five technician apprentices to Y-12: R.H. Williams, C. E. Hagy, T. G. Evenson, W. N. Tillery, and E. G. Stevens. Each was reminded to turn in their expendable items to the Tool Room by Harrill in one memo. J. D. Blanton and P. W. Hill received a copy of the memo indicating their role in management of maintenance at the time. A strong social and competitive atmosphere dominated the early years of operation at all the Oak Ridge plants. Softball was a major free time activity and the I&C fellows were strong competitors. Plant first aid teams were also popular. The I&C team won in a regional competition sponsored by the US Bureau of Mines.

Here Clarence Larson, president of Union Carbide presents a First Aid trophy to an Instrument Mechanics Team. Left to right: George Hamilton, Gene Tipton, Robert Splittgerber and son, James Smith, Gerald Hamby and James Day with C. E. Larson

John Blanton recalls that Bill Ragan was in charge of instrument maintenance in buildings 105, 106, 101-B and 205. Charlie Maxwell's group was outside the Q clearance area consisting of building 101-D, 104-B and 719-A. He says it took about three months to receive a security clearance, receiving his around the first of March 1948. Coincidentally, Union Carbide took over operation of ORNL in January that year and the Taft-Hartley law was put into effect. Blanton was assigned to the 717-B shop under Ken Forestall, foreman. By this time a new position was filled, that of quality control technician. Al Williams was the first hourly to fill such a position. About this time frame Gill C. Goss was assigned to be the Electronic Engineer for the group, Jack Davidson later replaced him.

The first apprentice class accommodated fifteen who began their training in 1948. Herb Linginfelter remembers being hired by Analytical Chemistry in 1948. At the time Byron Thompson was the "C" shift instrument man. His shop was in Building 105. Herb worked in building 205 initially, part of the Pilot Plant complex. Dr. Kelly, division director, frequently gave Byron written "Atta Boy" letters. Herb remembers Ocee. C. Cole being assigned to the D area down the hill. Herb signed a Bid List posted in his work area and came into I&C in late 1950. He reports being intimidated in his job interview by a man who looked, spoke and acted like Telly Savalis on the TV show Kojak. His name was Bill Ladniak. Herb was assigned to the maintenance shop in 3500 reporting to Bob Toucey. He remembers Tom Clabo, Gene Tipton, Jim Day, Bob Splittgerber, Charlie Kirkwood, George Hamilton, Emery Davis, Bob Davis, Ott Smith, Lorry Ruth, Howard Frazier, Joe Sutton and Byron who came off shift work. Howard Frazier "topped out" and transferred to the recorder shop along with Gene Hamby.

Joe Eddlemon remembers joining the 700 & 800 area shop in 1945 under the leadership of Al Simpson. In 1947 Al left to return to full time operation of his public address system business. Al Novak replaced Simpson as supervising engineer. Simpson had a maxim; My business is sound. When the war impacted his business he went to work at X-10. In 1946, he resigned and returned to his shop in Knoxville. Eddlemon left Monsanto to complete his education at the University of Tennessee. He had planned to study Electrical Engineering, but switched to Engineering Physics when he learned that high frequency was 62.5 cycles in that department. In 1948 Eddlemon was rehired by Cas Borkowski as an Instrument Technician.

Dave Cardwell who was Division Director of Engineering and Maintenance Division hired John Blanton in July of 1948. At that time, Bill Ladniak was head of Service Branch in the instrument Department. Monsanto was brought in at war's end to increase staffing and prepare for an expanded role for ORNL under AEC's new vision for National Laboratories. Blanton remembers that Al Williams was the quality control technician and Gill Goss was the Electronic Engineer. The release from military service of many who had been involved in radio and radar during the war helped ease the shortage of qualified personnel. Even so, it became necessary to establish an apprentice training program to meet the projected demand for maintenance strength.

Bill Ragan, supervisor; Gilbert Goss, engineer; Don Walker, mechanic, (shown R to L, all with the ORNL Instrument Department), are in charge of the project for operating the equipment to be used in the transmission of electric current for opening the gates to the Oak Ridge Area.

Quoting from a March 18, 1949 article in the ORNL News:

"Atomic power produced at Oak Ridge National Laboratory will figure spectacularly in the ceremony for opening the gates to the Oak Ridge area tomorrow. The official act for symbolizing the occasion will be the cutting of a ribbon stretched across the highway at Elza Gate at 8:30 A.M. It is planned to sever the ribbon by having an electrical current ignite magnesium a metal that burns with an intensely bright flame that will be contained in it. The current to be used is to have its origin in an electrical impulse generated by atomic power produced in the ORNL atomic pile. The impulse is to be sent by telephone connection between the Pile and Elza Gate The means for performing this historic gesture is described briefly as follows: Two telephone lines will be kept open between Elza Gate and the Pile. One of the groups responsible for the project will be stationed at Elza Gate at the time the ceremony is to take place. When the time comes for the cutting of the ribbon, he will, at a given signal, make known this fact over one of the telephone lines held open to one of his group situated at the Pile. Whereupon, the Operations Division personnel will be given the order to start operations in the Pile. As the power of the Pile reaches a certain elevation, it will cause an electrical impulse in an ionization chamber inserted in the Pile. The impulse will be carried over the other open telephone line to actuate a relay placed at Elza Gate. The ionization chamber to be used in connection with the production of power for the ceremony is in itself an important development in the history of atomic energy. Its great utility lies in the fact that it makes possible accurate measurement of the atomic power produced in reactors. The photo depicts three I&C maintenance staff members who prepared and executed the ribbon cutting by a nuclear signal generated in the Graphite Reactor.

Assigning this very public chore of creating and leading the Oak Ridge city gate opening was indicative of the stature I&C held at ORNL."

Organizationally, ORNL was gradually transitioning toward a more research-oriented organization with research-oriented groups and maintenance oriented groups tending toward separate organizations. At the same time, high level decisions within AEC and pressure from organized labor resulted in the formation of the Atomic Trades and Labor Council, a consortium of 16 American Federation of Labor craft unions, after an election held by the U. S, Department of Labor at the X-10 site. The instrument maintenance personnel were designated as instrument technicians, mechanics, checkers, or apprentices in the International Brotherhood of Electricians, Local 760. This sort of organization was informally based on the Dupont model for a typical company facility using function naming or buildings and personnel grouped by unit or shop. A significant political event was the passage of the Taft-Hartley law requiring the benefits package to remain frozen until the union contract with Union Carbide was concluded. Hourly pay grades were instrument Technician, Instrument Mechanic and Instrument Mechanic Apprentice. Initially the Technician rating was reserved for those assigned to shift work. By 1951 the Instrument Department had expanded to 132 total employees, with a separate research branch and separate service branch.

There were 68 people in the service branch under Ladniak. The 1951 organization chart lists 10 people under Bill Ragan in the 3000 area, 10 under Bob Toucey in 3500 and 18 under Ken Forrestal in Instrument Construction. Frank C. Sims (12/47) was transferred from Pile Operator Helper to I&C in December 1951. Further evolution of the Instrument Department occurred as the I&C Division was formed in 1953. It was during this period that Bill Ladniak transferred out of the I&C organization and the Maintenance Department of the I&C division was headed, briefly by Ed Bettis and then by Bob Affel, under Charlie Harrell's Instrument Department of the I&C Division.

Instrument Technician candidates in the Apprentice Program were given credit for certain previous associated education and/or military experience, which enabled them to graduate earlier. Blanton was promoted to Instrument Mechanic after two years in the apprentice program, as an example of this benefit from previous experience or education. After promotion, Blanton continued to work in the same general area, but began to specialize in maintenance of leak detectors, Creep Laboratory instruments, spectrometry and equipment in building 101-D. Blanton remembers fabricating scalers, count rate meters and other equipment not available on the commercial market. In 1949 Blanton was transferred to Charlie Maxwell's group, which was responsible for service and repair of stationary and portable radiation detectors, Monitrons, Leak Detectors, Whole Body Counter, Metallurgy Division and the Steam Plant instruments. Blanton remembers being transferred to building 7503 to join the new work on Aircraft Reactor Experiment which he continued until its conclusion in the mid 1950ís. His next assignment was the non-Destructive Testing Laboratory in building 3019. Since this was a new undertaking for the group, he performed both construction and repair of associated laboratory instrumentation.

Blanton says that about 1953 the upgrade of K-25 through K-33 and the construction of the Paducah Gaseous Diffusion Plant and Savannah River Plant resulted in the need for large numbers of experienced nuclear instrument workers. Charlie Maxwell, Charlie Vail, Jack Boone and J. C. Maxwell and others transferred to Savanna River. John Frisbee was promoted to take over Charlie Maxwell's group.

Formal apprenticeship standards were first published by ORNL in August 1948. They were contained in a ten page mimeographed booklet signed by three company representatives: E. W. Parish, C. S. Harrill and B. G. Catron. The three union representatives were: T. E. Rush, J. T. Maples and F. C. Sims. It is interesting that the Engineering, Maintenance and Construction Division had 616 hourly in 1948 and the Instrument Department had 36 hourly, never the less, had 1/3 of the representation on the ORNL apprenticeship standards committee. This trend of placing a strong emphasis on the instrument support issues would continue throughout the years to follow.

In February 1953, with formation of the new Instrumentation and Controls Division, an expanded apprentice-training program was developed. The training program required trainees to complete three years of classroom work, multiple work assignments to acquire skills in all basic technician tasks. The instructor pool came primarily from I&C Division engineering staff, plus journey level hourly personnel. The apprentice training program would continue for more than ten years.

Chapter 3

Early Electronic Instrumentation Development

This Chapter Generally covers the years 1943-1953, from the beginnings of ORNL (Clinton Labs) to when the I&C Division was formed.

Most of the early instrument innovation took place because there was little instrumentation available to buy for the new atomic energy field in which no one outside the university research labs, had ever done any work. Now, there were hundreds of scientists and engineers, where there had been, at most dozens. Physicists were striving to understand new physical reactions. Chemists had scores of new chemical interactions to understand. There was suddenly a need to regard the health of the researchers, who sometimes seemed prone to disregard their own health – so the field of Health-Physics was formed. Thus, physicists and chemists invented many of the first nuclear-instruments.

Some of the pioneers in electronic instrument development were the following individuals:

Floyd Glass
Hugh Wilson
Ed Fairstein
Bob Dilworth
Walter Jordan
Cas Borkowski

Many of the new measurements had to do with electrical pulses, as opposed to the measurement of steady state or slowly changing electrical quantities. Hence, the ability of an electrical device to measure and respond quantitatively to electrical pulses became central. Pule amplifiers were invented. Pulse-height analysis became more and more sophisticated as the field of nuclear instrumentation grew from infancy.

Some of the pioneers in instrumentation were former physicists who had a knack for electronics. One of these was Walter Jordan, who invented one of the first pulse- amplifiers, called the A-1. Others (including P. R. Bell) quickly made improvements on the early designs and soon, the designs of the ORNL instruments (those that were not classified as secret) were taken up by industry, as the nation at large began to do more and more nuclear physics experimentation. Industry did this quickly and increasingly, as the designs of the instruments developed at ORNL were in the public domain, having been invented through the use of public funds. Early in the 1950's the instrument Department issued drawings of these so-called "Q" instruments (actually all of the Instrument drawings were designated as the "Q-Series") which were in great demand by commercial instrument manufacturers.

The earliest organization chart found for the Instrument Department lists the following as engineers, working for Frank Manning: Floyd Glass, Hugh Wilson, Earl Siddnam, Gerry James. This group eventually became known as "The Electronics Section" and was one of the larger groups in the I&C Division.

Some of the devices invented or developed by Borkowski's group in the Chemistry Division were further developed in this group (which became a part of the I&C Div) when the I&C Division was formed. Ed Fairstein, who had worked for Borkowski in the Chemistry Division continued his imaginative and inventive ways in the group that remained to support the Chemistry Division, until he left ORNL to found (with Frank Porter) the electronics firm of Fairport Instruments, possibly in 1955. Frank Porter and Ed Fairstein parted ways after a few months, and the company name was changed to Tennelec Co., with Ed Fairstein as its owner.

The early days of the Instrumentation and Controls Division there were individual groups for detector, technology, mechanical development, electronic innovation and process control technology, under the Instrument Department, with Reactor Controls and special Research support groups joined under the overall Division Directorship.

Chapter 4

The ORACLE, ORNL's First General Purpose Digital Computer

Ray Adams

Perspective of 1946, as seen today

Upon the formation of the Clinton Laboratories, in 1946 and even up to 1954, there were no digital computers, as we know them today. As the United States entered WW II, a few Universities began to build electronic computers, starting with the Whirlwind Machine at MIT, finished in 1943, and the ENIAC machine at the University of Pennsylvania, that was finished in about 1946. These machines employed large numbers of vacuum tubes and electro-mechanical relays. Other machines, at Harvard and at Princeton followed these. Readers interested in the history of computers are advised to search on the Internet, for "History of Computing".

Compared to today's computers these early efforts seem trivial. The computing capability of even the smallest hand-held scientific computer (calculator) today, has about the same "compute" capability as those early machines. Furthermore, the largest fastest (scientific) machines today have speed and memory capacity millions (even billions) of times greater than those early computers. A small laptop computer today has greater speed and memory capacity than the largest and fastest computer of the 1960's.

Formation of the Math Panel

In 1946, at the Clinton Laboratories, there was a small group operating Friden and Marchant mechanical calculators, pounding out calculations as fast as their fingers could fly. This group was in charge of Bob Coveyou, and included Betty Maskewitz, who later founded ORNL's Radiation Shielding Information Center. Alston Householder arrived upon this scene and about a year later, Research Director of ORNL, Alvin Weinberg, asked him to take charge of this Mathematics and Computing Section of the Physics Division.

Then after about a year, when 'customers' were coming to the group from all over the Laboratory, the group was given divisional status. However, Householder, seeing that the group was so small, elected not to call it a division, but called it the Mathematics Panel.

Shortly after the Clinton Labs became the Oak Ridge National Laboratory, in January 1948, researchers realized that computational tools were needed that surpassed the ability and were more accessible than the group of Friden and Marchant calculators. Researchers needed more than even the IBM card-programmed calculators that were being used (primarily for business purposes) at the K-25 and Y-12 plants could provide. Computational time was for awhile, obtained at the Mark I machine at Harvard, but the calculational needs of the projects then underway or envisioned by the engineers and scientists at ORNL, were deemed to require additional, and faster computing capability.

As the Director of the Math Panel, Householder turned from his earlier love of mathematical biology to numerical analysis. He became well known in numerical analysis circles. His publications on numerical analysis became central in the Mathematical/Computational arena. In 1949, NEPA (Nuclear Energy for the Propulsion of Aircraft) a project operated by the Fairchild Company (separate from ORNL) actually developed a special-purpose digital computer for the solution of linear equations. That machine was called Oak Ridge Automatic Computer for Linear Equations – ORACLE[7] After Fairchild left, some of the work was taken to ORNL and there it was called the ANP program. ANP and AEC requested that ORNL investigate the acquisition of a modern digital computer for the lab.

Argonne National Laboratory near Chicago, had formed a Digital Computer Development Group. The Argonne effort was led by Dr. Jeffrey C. Chu, who was one of the pioneers in digital computer design, along with several people at the Princeton University Institute for Advanced Studies, including J. Presper Eckert and John von Neumann. John von Neumann had been retained as a consultant to the ORNL Math Panel

A Computer – by Committee

An ORNL committee consisting of P. R. Bell, A. S. Householder, and Lewis Nelson, was charged with recommending a computer for ORNL. Following visits to RCA Labs, Princeton, the Eckert-Mauchly Computer Machinery Company, and with consideration given to Raytheon and Sylvania efforts, their study first recommended the purchase of a Raytheon machine, but because excellent progress was being made at Princeton, with a machine that used the Williams Tube fast memory, the committee reversed its decision and recommended that ORNL pursue an agreement with ANL, for the joint construction of a copy of the Princeton IAS machine[8]

ORNL86536

A. S. Householder, (shown with a later computer of which the ORACLE was the first of many at ORNL.

The successes of recently devised electronic computers, especially at the Princeton Institute for Advanced Studies (IAS)[9] indicated they were necessary tools for those working on such projects as the reactors for Aircraft Nuclear Propulsion, and even the general scientific computations of the ORNL Scientific staff.

We'll Build Our Own

The joint project, initiated with the Argonne National Laboratory in 1950, resulted in a contract with Argonne, by which both Argonne and ORNL would produce (at Argonne) copies of the Princeton Machine. Argonne's machine was called the AVIDAC (Argonne Version of the Institute's Digital Automatic Computer) The Oak Ridge machine was called ORACLE (Oak Ridge Automatic Computer and Logical Engine). Four engineers, newly hired for the purpose of fulfilling this ORNL need, were sent on-loan to Argonne.

Those engineers, sent to Chicago for this task, were Earl (Earl W) Burdette, Bill (William J.) Gerhard, Bud (Rudolph J) Klein, and Jim (James W) Woody. They left Oak Ridge in mid 1950, for the effort, scheduled to last 6 months. Organizationally, those four engineers were a part of the Engineering and Mechanical Division of ORNL. They worked in the Service Branch under Bill Ladniak, for Charlie Harrill, who was director of the Instrument Department, under Cardwell.

Shown in front of a part of the ORACLE
are L - R, Gerhard, Woody, Burdette & Klein

Rather shortly, the AVIDAC was produced, but for reasons not clearly identified, the ORACLE became a more advanced machine, which of course, took longer. The work under Chu at Argonne, was stimulating to the I&C engineers. One of the differences the ORACLE had was that whereas the IAS machine at Princeton did its arithmetic in a serial fashion, the ORACLE was designed to do operations in parallel - which made it much faster.

The effort originally planned for 6 months, stretched to almost 3 years, after which the ORACLE was moved to ORNL, for installation in the newly completed Bldg 4500, where Householder's Math Panel eagerly awaited its operational status. By the time the ORACLE was installed, the I&C Division had been formed, which included the Instrument Department, organizational home of the ORACLE's four development engineers, Burdette, Gerhard, Klein and Woody.

It Wouldn't Go in the Building

Installation of the ORACLE in Bldg 4500 involved physically removing parts of some outside walls and included the addition of 40 tons of Air-Conditioning to cool the machine's 5,000 vacuum tubes.

ORNLXXXX

The ORACLE Being Moved Into Bldg 4500

Shortly after the ORACLE was installed at ORNL, the I&C engineers found that cathode-ray tubes used for the Williams tube memory were improved. Bud Klein made an in-depth study of a new industrial type of cathode-ray tube, and recalls that feedback was designed-in, to stabilize the beam-current. In a report issued July 31, 1954, he reports that other improvements were made as well, to the end that the memory of the machine could be doubled, to 2048 words. This result occurred in about March of 1955, recalls Jerry Sullivan, who had recently come to ORNL, to work in the Applied Neutron Physics group.

As users of the ORACLE for their calculations, Bob Coveyou's group (which included Jerry Sullivan) wrote a "compiler" for the machine, that greatly simplified the inclusion of subroutines and the various constants required for most calculations. One could, by using the compiler, include the benefits of a much larger body of codes that other people had written and tested.

ORNL12304

Bud Klein of the I&C Division shown at the console.

The Women Mathematicians

The Math Panel employed a number of women mathematicians, who set about writing machine code for many of the subroutines, as well as doing programming for ORNL's quickly increasing calculational needs. I recall that there was a quick acceptance of Mozell Rankin, Susie Atta, Nancy Dismuke, Arlene Culkowski, Nancy Alexander, Jeannie Harrison, Nancy Betz, and others, who were equally quick to help out us "duffer" engineers, seeking to write code for the ORACLE.

Some of these women earlier went to Argonne to work on programming there. Mozelle Rankin recalls that she wrote a sine subroutine using the MacLaurin expansion during the summer that she, Virginia Klema, and Ruth Arnette spent there. The employees newsletter of ORNL featured these women mathematicians in an article that appeared in the ORNL News, June 16, 1961. A photo of several of the women mathematicians is reproduced here.

ORNLNews83XX

left-to-right are: Jeuel LaTorre, Barbara Flores, Marjorie Leitzke, Margaret Emmett, Nancy Betz, Arline Culkowski, Joan Rayburn, and Nancy Alexander.

As Director of the Mathematics Panel, Alston Householder established a rigorous set of standards for approved subroutines. However, the standards were so tight that many authors of subroutines chose to test their work to a lesser standard that most users thought was "good enough" for their calculations. Consequently, the body of "good enough" but very useful subroutines soon grew to be substantially larger than those that were rigorously "approved."

Operation of the computer was under the able direction of Charlie Williams. Two of the operators were Sherril O. Smith and Hollis Stakes. The operators were well trained and skillful at their tasks. Input to the computer was via punched paper tape (Prepared off-line) which was fed by the operators into a 300 frames-per-second tape reader, after which the operator would start the computer to begin operation of the just-read program. The command to start was "4 3 to zero and Op".

An occasional user/programmer of the ORACLE, I was at first puzzled by the disclaimers associated with the larger body of subroutines. They sometimes said something like "This subroutine has not been exhaustively tested so as to make certain its results are true, beyond doubt. However, the user may use it for calculations the results of which should be subject to careful scrutiny." I quickly found that none of my calculations needed the "higher" standard.

World's Fastest

About the time the ORACLE arrived in Oak Ridge, in the Fall of 1953, an Argonne press release proclaimed the ORACLE the "World's fastest high-speed general purpose digital computer ... with the greatest capacity of any yet built." As it turned out, this accolade endured for only a few weeks, before IBM began to deliver its model 701 computer the 2nd quarter of 1954. The IBM 701 also used the Williams tube fast memory.

By the time I arrived at ORNL, at the beginning of 1954, the engineering group had come back from their stint at Argonne. I asked one of them what contributions their leader, Earl Burdette, had made. Bud Klein said, "Oh, Earl was very good at keeping the administrators out of our way, while we made the needed engineering innovations."

As the ORACLE was installed and tested in its new home, the I&C Bargaining unit employees - we called them technicians, quickly became proficient in the newly established use of vacuum tubes as digital logic units. That early group of Instrument Technicians included L. R. (Ralph) Gitgood, J. A. Graves, Earl McDaniel, and Dave Ramsay. Later, Burt Denning joined the group. Over the course of its use at ORNL, the I&C engineers and technicians made numerous improvements in the ORACLE.

Among the first improvements were a fast paper tape reader, that would read 200 or 300 7-bit frames per second, and a paper tape punch that would punch 60 characters per second (cps). One technician in particular, Ralph Gitgood, seemed to have an uncanny knack of making a 60 cps tape punch work at full speed. These improvements greatly enhanced the I/O speed with which programs could be input to the computer, and the rate at which results could be output. Later, another device was added, which greatly speeded up computer output. This was the ORACLE Curve plotter[10]

ORACLE Curve Plotter

The curve plotter was initially envisioned as simply, an x-y graphic means to plot the results of ORACLE calculations, and it did that (see below). However, it had the ability to quickly draw densely spaced numerical, and alphabetic characters, (a matrix of 30 x 51 characters in about a second) and had a fast camera to take pictures of the output. Each film magazine of the device would hold 200 pictures. The curve plotter was engineered as an addition to the ORACLE, by several of the members of the I&C engineering team, which by 1959 included, Bud Klein, Marilyn Chester, Jay Reynolds, Carl Schalbe and Jim Woody. It was quickly employed for alpha/numeric "printout" of computed results and even hexadecimal memory dumps. In a month of ordinary curve-plotter use, about 5000 pictures would be made. It was this output mode that made the ORACLE attractive to the ORNL Budget office, for their monthly reports.

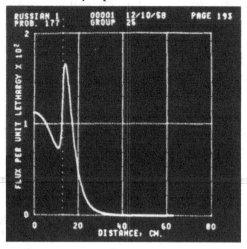

ORNLCP1258

Curve Plotter output of Graphical Results

ORNLCPPOG

Curve Plotter output of Text

Operational Maturity

A "sociological" event occurred in the Math Panel, about this time. The question as to whether or not "outsiders" (people outside the Math Panel) would be permitted to "run programs" on the ORACLE, or should the problems that the machine worked on, be entered and controlled only by members of the Math Panel hierarchy. The ORACLE operations group rather strongly asserted that they should be the only ones to "run" problems on the ORACLE. I do not know how the controversy played out internally, only that the head of the operations group was dismissed rudely by the posting of a memo on the bulletin board - and henceforth anyone could run programs on the ORACLE. Although I agreed with the final "open-shop" decision, I (and others) thought the mode of the dismissal of the operations chief, to be shameful!

The I&C Technicians worked hand in hand with the engineers to make the ORACLE one of the most reliable machines of the vacuum tube era. I've already mentioned Ralph Gitgood and his ability to make a paper tape punch work up to the 60 cps design speed. Ralph was a veritable mechanical genius who also helped keep the many automatic typewriters working that were kept in a room alongside the ORACLE machine room. Earl Mc Daniel, Burt Denning, Ralph Gitgood, Dave Ramsay, and others were quick to fix any problems that the engineers identified, and were equally quick to make the many improvements that took place over the computational life of the machine. One of the facts of life of that (or any) vacuum tube machine, especially one that had thousands of vacuum tubes, was that the tubes had a tendency to "go to sleep" and not to "wake up" when their digital status changed. That and other problems with the tubes were lessened by pre-aging the tubes in a warm-up rack that held perhaps, as many as 1000 tubes in readiness for their placement in the machine. One interesting anecdote, provided by Bill Busing, a Chemist who did many ORACLE calculations, is related as follows:

> "My ORACLE program stopped after trying to take the square root of a negative number. My dump showed that I had gotten the argument by dividing one negative number into another negative one so the result should have been positive. When I complained, an I&C engineer or technician suggested that I put that division into a loop while he put a scope onto the divider unit. And pretty soon he said, "Oh yes, there it is." Whereupon, he changed a tube and all was well. Those were the days!"

The ORACLE provided 8 years of work-horse service at ORNL. It was retired in the Fall of 1962, at which time a Control Data Company Model 1604 was installed, the first of several com-commercially produced machines to be put into operation at ORNL.

Chapter 5

Maintenance 1953 & Beyond

Don Miller

It was an exciting time for the Instrumentation and Controls Division (I&C). With the memory of being in charge of the gate opening for the city in 1949, the formation of the division in 1953, and having a larger role in the greatly expanded research and development programs, I&C maintenance was a happening place! The architect's photo of the proposed two-story addition to building 3500 that was published in the Laboratory News is shown below.

ORNLBldg3500

1955 proposed addition to Bldg 3500. The addition was completed in 1960.

John Blanton says a topic of importance to the Instrument Mechanics at the time Ed Bettis took over I&C Maintenance supervision vacated by Bill Ladniak, was pay brackets. Since the failed Monsanto plan to raise hourly pay by twenty-five cents, pay brackets had been frequently discussed. Finally, Charlie Harrill, Ed Bettis, the supervisors and central compensation agreed to eliminate the mechanic position and make everyone either an apprentice or an Instrument Technician (I.T.). At the same time, a classification called Red line Instrument Technician was created. This classification was paid slightly more than an I.T. and was viewed as a subject matter expert and task leader within the hourly ranks. The position was qualified through compensation by being paid the premium only on second or third shift. Virginia Farris (a long time I&C Div. Secretary) remembers a few Instrument Technicians who eagerly sought her out on Mondays to see who recorded overtime for work performed over the previous weekend. A spirited competition existed to ensure that equalization of overtime pay within the I.T. overtime list was maintained. The Red Line classification was phased out by a decision in Labor Relations, it is believed. The last Red Line disappeared about 1980.

Joe Eddlemon remembers "Yes, Al Novak was the group leader when I worked in the Instrument Department in 1947. When I joined the Department in December, 1945,

Al Simpson was the group supervisor. He had a business in Knoxville; his maxim was "My business is sound." He rented public address systems for various occasions. When the war impacted his business, he went to work at X-10. In 1946, he resigned and returned to his shop in Knoxville. At a party at his home, I remember Bob Toucey, Paul Hill, Phil Williams, and a fellow from Louisiana whose name I cannot recall. I now remember J. R. Jones in the group. Page Buckley was an engineer in the group. Paul Hill, Phil Williams, O. C. Cole, Bob Pratt (not certain about this name), Jack Boone, and I were in the group. Our shop was in Building 706A. When Al Simpson resigned, Al Novak took over and was in charge when I took leave from Monsanto Chemical Company to attend the University of Tennessee. I planned to study Electrical Engineering, but when I learned that high frequency in the department was 62.5 cycles (hertz), I switched to Engineering Physics. In my sophomore year, management of ORNL changed to Union Carbide Corporation from Monsanto Chemical Company. In the middle of my second quarter, Union Carbide notified me to return immediately or be terminated (I presume that is a term considered to be kinder than being fired). So I was fired.

Casimer J. Borkowski's group. As a technician in the Instrument Group, I had done In the fall of 1948, I was rehired by ORNL and worked in maintenance on the instruments in his laboratory and seemed to like the work I had done."

[Herb] Linginfelter recalls "You asked me to participate in writing a history of the now defunct I & C Division at the Oak Ridge National Laboratory. Here 'tis. It is difficult to even think of I&C division without having my own life woven into the account, after all, I grew up there. Retiring at age 53 and going into construction as a Job Site Manager, I found I really missed the people at "the Lab". This gives me a good opportunity to think back to the good old days. Knowing I can't begin to think of all the men and women who made up I&C, I will give it my best shot to say a word about those few with whom I came in close contact. In the 1940's, I&C Division had the responsibility for the design and construction of instruments which did not exist anywhere in the world. The nuclear age brought about this necessity.

Engineers such as Roland Abele, Tom Gayle, Tom Hutton, Harry Todd, Phil Williams, George Holt, Wayne Johnson, Wiley Johnston, Bernie Lieberman and many others came through with designs of instruments that could detect radiation, that could count the DPM's (disintigrations per minute), nowhere else could this be done. Even our division director, Cas Borkowski, came out with a design for a Personnel Radiation Monitor, a device so small it was worn in your shirt pocket and gave an audible alarm when radiation was present, it even read the amount. Instruments that could identify chemical elements that could tell the constituents of any form of material. I'm getting carried away in the moment.

Hiring into Analytical Chemistry Division in 1948, I remember Byron Thompson as "C" shift's instrument man. His shop was in Bldg. 105, the reactor building. The lab where I worked was in Bldg. 205, part of the old Pilot Plant, and Byron took care of all the Instruments in this lab as well as our counting room and other shift facilities on the hill. Dr. Myron Kelly was director of Analytical Chem at this time and gave B.C. lots of "atta boy" letters. O.C. Cole was the tech down the hill around Bldg 706A and the "D" Bldg area. I signed a "bid list" and came into I/C in late 1950. I consider this the best luck I ever had. I was intimidated in a job interview by a man who looked, spoke and acted like Telly Savalis on the TV show "Kojak". His name was Bill Ladniak who, I think, was in charge of all maintenance. The maintenance shop I was assigned was in 3500 Bldg and Bob Toucey was supt. Some of the men in that shop at that time and for a few years on were Tom Clabo, Gene Tipton, Jim Day, Bob Splittgerber, Charley Kirkland, George Hamilton, Emery Davis, Bob Thomas, Ott Smith, Lorry Ruth, Howard Frazier, Joe Sutton, and B.C. Thompson who Came off shift work.

Paul Hill became supt. of this shop when Toucey left. Paul was promoted to Maintenance Supt when Charlie Harrill left and Jim Day became supt of this shop. Howard Frazier topped-out and went into the recorder shop along with Gene Hamby. Charlie Kirkwood, Joe Sutton and Ott Smith went to Saudi Arabia to work for Aramco Petroleum and never came back to the Lab., hoping to make a go of it in their own business. At one time, Wiley Johnston was in charge of the "Standards Lab"in the back end of 3500 Bldg. I believe George Ritscher was over that lab earlier. I do remember NO ONE was allowed into that lab without the superintendent escorting you. It was the inner sanctum, you couldn't even see into it. I always tried to sneak a peek through the doorway if someone came in or out as I walked down the hall. Never did. So imagine what I was thinking when one day at lunch while most of the guys from the shop were out back with a football, I kicked it through one of the back windows of that lab. I figured firing me would be an easy way out. It would beat blindfolding me against the brick wall and giving a rifle to Wiley. But neither happened. I was advised to not let that happen again. It didn't.

Lorry Ruth and I decided to lose weight. We quit shoveling in do-nuts at the cafeteria for breakfast and brought to the shop large size boxes of corn flakes, sugar, milk and bananas. In two weeks time, we had eaten 3 large boxes of cereal, over 2 pounds of sugar, almost 7 pounds of bananas and washed it down with about 2 gallons of milk. The nurse at the health dept told us it would be better to stop our diet after we weighed in 6 pounds heavier. We did. But somehow we managed to continue to gain weight.

When I retired in April, 1981, Paul Hill was Supt. of all Maintenance personnel. Charlie Mossman was over all Field Engineering and Frank Manning was over R&D engineers. Cas Borkowski was division director. I was responsible for 2 repair shops in 4500S, 1 shop in 4500N and a shop in 3019 bldg. Roy McKinney in Manning's group was set up as foreman over the 4500N shop as I was mustering out.

Highlights, as I remember them: Placing Paul Hill in charge of all maintenance shops. This put all union problems, including the 100+ grievances under one man. All shop foremen and all instrument technicians reported to one man. Paul Hill did an excellent job. Shifting engineers into Moss man's or Manning's groups. This put help for instrument shops in direct organizational alignment with the design engineer. Low points: Deciding to abolish apprenticeship training thinking well-trained people could be hired outside the Lab. In fact, the best technicians we ever had were the topped-out apprentices. Several applicants "snowed" our interviewers and quite a few people were hired who just didn't fit in. I could go on and on but I won't.

I do want to list the technicians who were in the 4500S Analytical Chemistry shop at some time during my tenure. They were the best of the best - Gary Laxton, Troy Chambers, Ben Carpenter, Larry Lane and Brent Davis. Howard Enix was the tech in 3019. All of these were excellent technicians, excellent character and, by the way, all of them were "home schooled" in the I&C apprenticeship training school. I hope some of this will help shed light on our I&C Division folks and maybe even show some of the pride I think most of us had by being in I&C."

John Blanton remembers the extra tasks that fell to I&C maintenance because of the can do and innovative approach they had demonstrated in the early years. All ORNL audio-visual needs, two-way radio systems, intercom systems in nuclear facilities, paging and personal radiation monitors were generally purchased or designed by I&C and maintained by the maintenance department. As monitoring radioactive waste streams and air grew to a major program, I&C maintenance took the lead in supporting this work. This included lakes, streams, storage tanks, stacks and reactors which required reliable and accurate monitoring and trending of radiation, effluent volume and noble gases. New equipment developed and manufactured in-house required operating and maintenance manuals. I&C engineers prepared these manuals for use by maintenance and end users. These expanded duties required additional hourly staff and further specialization of individuals and shops: Communications and Security, Reactor Systems, Special Electronics, Mainframe Computer, and Terminal and Personal Computer maintenance to name a few. Also some tasks were so specialized that a few Instrument Technicians were assigned to other divisions as individuals who worked only on highly specialized projects, such as Environmental Science and Isotope Production.

As the numbers of instruments multiplied with the advancement of existing and new programs, the creators of the various instruments in other divisions were happy to relinquish responsibility for maintenance to I&C. This was especially true when other projects and divisions within the site adopted the use of their designs and left responsibility for support to the instrument people. Also during this era, the specialization continued an earlier practice of having hourly shops in the engineering chain of command, thus assuring a close link between design, fabrication and installation of new equipment not available off the shelf outside the Laboratory. The term Cradle to Grave came into common use in an attempt to describe the Division's ability to create a new design, manufacture, test, install it and provide maintenance until it was ready for retirement. A major early effort was in the design of radiation survey instruments as shown here. It is a Personal Radiation Monitor designed by P. R. Bell in 1946 and it was fabricated in instrumentation shops.

Editor's Note: P. R. Bell was a gifted innovator of instrumentation during his long tenure at ORNL. He was considered early on, when the I&C Division was formed, as a possible Director of the I&C Division. I sent the picture (reproduced below) to Bob Dilworth who wrote Chapter 7 of this book that describes a more modern version of a "Personal Radiation Monitor." Here Bob's reply:

I don't recall a PRM by P. R. Bell - who I did know well. The 1946 date of his work predates my arrival at ORNL in about 1956.

However, it looks much like a Victoreen electrometer tube and a sensitive relay and perhaps an ion chamber.

If so, that is the configuration of Floyd Glass's R-Vox instrument, which would squeal when a set value of accumulated dose was reached. The center electrode of the ion chamber had no resistance to ground, and connected to the electrometer tube grid. Therefore it required extremely good insulation so that leakage would not discharge the accumulated voltage. Floyd knew every trick in the book about very high resistance insulators. He also designed a resistance bridge to measure the values of the very-high-megohm glass-tube resistors that Victoreen used to make and which were used in electrometer circuits. When I was at Spinlab in Knoxville before coming to Oak Ridge, we made and sold a few of those bridges, and learned how to use freshly machined polystyrene, baked for a while, and coated with "Q-Dope". The R-Vox was very difficult to manufacture, and alarmed on an accumulated dose, without telling you where you got it. After the criticality accident at Y-12 when people didn't know which way to run, Cas was insistent on getting a dose-rate sensitive personal alarm - hence my PRM.

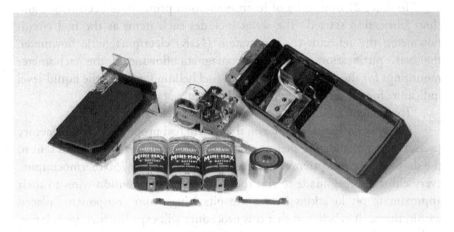

ORNL8641

Caption: Very early version of a "Personal Radiation Monitor" that was fabricated in the instrumentation shops.

The ANP Project

A reactor division quarterly report provides some details about I.T. work on the Aircraft Nuclear Propulsion (ANP) project. "The reactor design, with the exception of very minor changes, remains as previously described. All detailed drawings have been released to the shop. The beryllium oxide blocks have been sized and are ready for assembly into the core. The pressure shell has been received from the fabricator (Lukenweld). The reinforcing segments for the head have been welded in place, and the head holes have been bored.

The various core pressure-shell components are currently being fabricated and assembled."

R. G. Affel, of the ANP Division reported that installation of all instrument panels had been completed. "Approximately 80% of the process instruments were mounted and supplied with electric power and/or compressed air in the previous quarter. The remaining instruments should be installed by January 15, 1953. Minor instrumentation changes have been made to accommodate the design changes in the pump seal. All major process instrumentation components were detailed and either on hand or on order. It was planned to complete the installation of instruments on the panels so that as work in the pits proceeded, the sensing elements, when installed, may then be checked directly to the installed panels."

"To date, 21 of a series of instrumentation prints have been issued and shop fabrication started. The series includes such items as the fuel circuit flowmeter, the reflector-cooling system (NaK) electromagnetic flowmeter, the NaK purification system electromagnetic flowmeter, the tachometer mountings for the reflector coolant and fuel helium fans, and the liquid-level indicators for the reflector coolant and fuel surge tanks."
*The Lukenweld Division of The Lukens Steel Co."
"Location of thermocouples on the system had proceeded at a satisfactory pace. Two prints showing thermocouple construction were issued and sent to the shops. Since the system would require approximately 750 thermocouples, every effort is being made to run the thermocouple extension wires to their approximate pit locations before the pits have major components placed within them. It is believed that this procedure will expedite final installation and test of the temperature instrumentation."

Thermocouples were used by the thousands in many different reactor projects over the years at ORNL. I.T.s were required to know and understand the different types of thermocouples and have skills in their proper installation and connection to recording devices.

Instrument Checker Category

The following entry in a 1968 seniority list of hourly workers has a special significance. During the early years of operation in house designs were usually one of a kind pieces of equipment and were evaluated for correct operation by their designers following their assembly. However, for a brief decade between 1960 and 1970, equipment "manufactured" in-house grew to such a large number that a special category of hourly worker was created.

INSTRUMENT CHECKER GROUP

POS	BADGE	NAME	SENIORITY DATE
001	01827	BURGNER, BEATRICE R.	01-04-44
002	17098	FORD, BENJAMIN A.	09-25-67

These specialists were trained to make specific tests of electronic equipment following completion of their manufacture. Equipment such as Geiger counters, area radiation monitors, air-flow monitors and pulse counters for use in reactors were built by the dozen at times. Lengthy procedures and extensive test data was collected and stored for years as part of the pedigree of the equipment.

1966 to 1983 I&C Maintenance

Editor's Note: This section contains several "personal anecdotes" of Instrument Technicians, as well as statements about the field shop organizations of the Instrument Technicians.

The late 60s through the early 80s, I&C Maintenance was marked by a robust innovative spirit in design and fabrication of unique instruments and controls at ORNL. Electronic computers came of age and technology breakthroughs in sensor and microchip technology permitted a host of new applications for research and development for other Laboratory programs. Maintenance shops grew in numbers, recruiting staff from the volatile aerospace and computer industries as well as individuals leaving the military. Major programs with names like; MSRE, HFIR and Chemical Technology Division's pilot plants demanded the best from I&C.

A May 1977 I &C organization chart lists five **Instrument Technicians**; R. A. Francis, C. P. Littleton, D. C. Livingston, J. N. Smith, B. E. Vanhorn assigned to Operating Reactors under the technical leadership of three Engineers; D. S. Asquith, J. B. Ruble, D. D. Walker and a craft supervisor: J. M. Farmer.

These and other field shops required a solid pipeline of new equipment designs and redesign of existing instruments. A unique engineering group; The (analog) Circuit Development Group under the technical leadership of H. N. Wilson provided many of these new applications and inventions. In close partnership with the design team was a small but important group of Instrument Technicians: J. H. Burkhardt and C. T. Stansberry. Stepping into the future with (digital) applications was Computer and Pulse Techniques under the leadership of J. W. Woody. This design group evolved into a major business within the Division. Applications of computer-like devices and computers to every phase of development work at ORNL was their job. Instrument Technicians T. W. Ayers, R. J. E. Bradford, R. M. Childs, B. A. Denning, D. R. Dunn, E. L. Glandon, D. L. Holtzclaw, C. E. Houston, J. H. Johnson, A. J. Millet, C. R. Mitchell, D. G. Prater, R. P. Rosenbaum and B. A. Tye were supported by Engineering Assistants C. R. Cinnamon, L. R. Gitgood, C. C. Johnson and C. W. Kunselman. This large group of specialized maintenance staff was ably led by J. A. Keathley General Supervisor.

Editor's Note: The various instrument shops were employed to fabricate racks or panels of instruments for various jobs. Shown below are some examples of the instruments and instrument panels that were fabricated in the various shops. The recollections of the Instrument Technicians (I.T's) are also given here, as recalled by Don Miller. Their recall deals less with the hardware they maintained and fabricated than with the more "colorful" aspects of their human interactions.

Lorry (Shucker) Ruth recalls "I came into the instrument department in 1951 from [being a] reactor operator and spent a year in the circuit shop then in 7503 building (ANP project). John Blanton and Conway Moore were my co workers and Bob Affel was the project manager.

Admiral Rickover came to visit [one day] and Ed Bettis was standing around the guard rails in building 7503. I was in the pit working on valves. Ed told the admiral if they let water get to the blanket "we've ploughed up a snake." And the admiral looked at Ed like he did not know what Ed was talking about.

George Keller stories:

I was working in the valve shop in 3550 when I killed a small garden snake. I had an empty candy tin. I put the snake in the tin and put a wire around its head and taped its head to the lid of the tin. I took the tin to the electronic store and set it on the counter. George wanted to know what was in the tin. I told him it was candy but he would have to taste it first. He opened the tin lid saw the snake and threw the tin about 12 feet. He proceeded to call me every thing BUT a child of god!

When the time clock was at the west portal, George and I were walking down to the clock-out building. At this time [Frank] Rau came tearing down the hill on his bike and took a hard fall. George came out of the time building, crossed his arms and said "You're Safe". George Hamilton, who was there, said that Frank got up, rather sore from the hard fall in the gravel, and made some upcomplimentary remarks to George Kellar, who said something like, "I could have called you Out, Frank".

Splittgerber stories:

When I was at 7500, [R. B.] Splittgerber came in the building to tell this story. He bought a car from his brother in law in Clinton with cruse control. He kept running off the road with his new used car and could not figure out what was wrong. He thought the cruse control was supposed to guide the car.

Splittgerber came in one day all scratched up. I asked him what happened and he said he wanted to give his cat a bath but the cat would not get into the water. Splittgerber decided to get into the shower and get the cat wet. The cat Won. No Bath!"

This story is typical as Linginfelter remembers it. "Splitt" was assigned a work order from Chemical Technology Division to mount a Brown Recorder in the top of a relay rack along with other panels and deliver it to Building 4505. During the transit and delivery phase of the job, Joe Wallace drove the truck. "Splitt" rode in the back to guard his handiwork and protect it from bouncing too much. It was a long story as to how it happened, so let's just say a 12-inch wrench fell from Splitt's hand and broke all the glass out of the recorder door. Ashamed, Splitt had Joe return him and his mess to 3500 Building for repair. After replacing the door glass, they once again set out for 4505 Building.

They took the rack up the elevator to the second floor where Splitt began to roll it through the building looking for the customer. The floor was sloped toward a large plug in the floor, maybe 12-14 feet across used to lift large equipment by crane to the second floor. Splitt did not know that the plug had been removed that morning, and the slope caused the rack to roll faster than he could keep up, and the momentum of the rack prevented him from stopping it. The rack, with its two lead bricks in the bottom for ballast, plus the recorder and all the remaining instruments, dropped like a shot through the floor plug opening and Splitt came close to following it down. As the dust and debris began to settle, I thought to myself it was good that Splitt let it go when he did. The fall demolished all.

Stories happening in a distant instrument shop:

Linginfelter remembers that quite a few incidents occurred in shops where fewer prying eyes were present. "Howard Frazier got the idea from somewhere to use his bench-mounted Variac and two stainless steel welding rods to cook wieners for lunch. He adjusted his Variac down to "Zero" volts on its scale, turned its supply voltage "Off" and then stuck a welding rod in each end of the wiener. When he turned the supply voltage back "On", he would adjust his Variac slowly upscale. You could see the wiener begin to "cook" as the voltage increased and within 10-15 seconds, he had a beautifully cooked hot dog – delicious. Later, George Hamilton took the top cover off Howard's workbench and reversed the two output wires from the Variac.

We all gathered around Howard's bench at the next lunch, acting as if we were waiting our turn to cook on his marvelous hook-up, but actually we were waiting to see what happened. With a smug look of expectant success, Howard started through his proven routine, appreciating the large crowd around him. When he turned the supply power back on to the Variac with its scale set to Zero, little did he know he was putting 135 volts instantly across the wiener. There was a loud explosion followed by hot grease and wiener fragments flying everywhere. Frazier spent lunch-time cleaning that side of the shop but all the rest of us really enjoyed the show except Supervisor, Bob Toucey." Because of Howard's experience, that was another thing that wasn't done again. A book could be written about all the shenanigans that happened over the years.

TWO MEN/ONE ACCESS

The Oak Ridge Research Reactor (ORR) was the first to produce a high-level neutron flux for multi-experimental purposes. It was originally designed for a maximum power level of 10 Megawatts, but a higher level of flux soon became desirable. The reactor vessel fuel, and control features were designed for an increase of power to 20 MW. However, the Balance of Plant (BOP) had limited water-to-air heat exchanger capacity. Water-to-water heat exchangers and a secondary Water-to-air cooling tower were required. These changes warranted many modifications in the I&C portion of the BOP systems.

The reactor coolant of demineralized water, at a maximum flow rate of 20,000 gal/min. was controlled by a 24" butterfly valve. The pneumatically controlled butterfly valve was located in a below ground concrete lined pit and shielded with a concrete cover containing a single 20" manhole access. The valve position was automatically adjusted for the reactor heat being generated. Thus the position was critical to successful reactor operation and stability. The valve pneumatic controller was located in the main reactor control room and it was imperative to have accurate alignment of the valve position and the control room position designation.

To that end and before reactor startup, Foreman Raymond Tucker and Instrument Technician Lory Ruth were in the valve pit observing the valve position and transmitting same to the control room. Little did we know the characteristic of large butterfly valves when close to being full shut, i.e., the water flow forces on the valve plate tend to reverse and the valve will slam shut which initiates a water hammer. The hammer is shocking to say the least, when some portion of 20,000 gpm suddenly stops. It happened and both Foreman and Technician decided to leave the valve pit simultaneously through the only 20" egress. Whatever the case of who exited first no one will ever know but the first thing observed was the two standing on the valve pit concrete cover pad congratulating each other.

And finally Linginfelter remembers "George Hamilton bet me $1 I couldn't carry a 96 pound micro-manometer to the third floor of the thorium building, 3508, without sitting down on the stairs and taking a rest. After much huffing and puffing I finally made it. He paid me the dollar and grinned at me. Then I knew I had been "had". It was George who needed the manometer to test a pressure transmitter and he had suckered me into getting it up on the third floor for him. Fifty something years have passed but he still reminds me of the "Huck Finn's fence whitewashing" he gave me!"

ORNL 93187-68

John Frisbie's shop fabricated this Recorder/Variac Panel for the HFIR

The Special Electronic Support Group under J. L. Lovvorn included both design and fabrication capability. Instrument Technicians: H. L. Barnawell, J. L. Basler, M. S. Blair, W. A. Bratten, R. H. Brown, E. D. Carroll, J. D. Culver, E. A. Davis, B. L. Dennis, H. T. Enix, J. H. Fairs, D. L. Foust, J. A. Goan, R. L. Green, J. D. Harrell, F. E. Hatfield, J. R. Hendrix, N. B. Hickman, G. D. Inman, C. R. Ketner, G. W. Kwiecien, S. F. Lanthorn, J. W. Lawson, C. M. Malone, W. T. Martin, C. R. McAmis, J. E. McCarter, N. W. McCoy, C. R. Moree, T. N. Muncy, E. G. Price, F. M. Rau, W. T. Roberts, E. G. Rose, R. E. Saxton, J. H. Sherrod, R. S. Thomas, and H. H. Tompkins were engaged in manufacturing of a wide variety of instruments and controls. This activity was subdivided into groups who built prototype units for testing, custom units for special applications and routine fabrication for use in nuclear reactors, ORNL research divisions and by Health Physics in radiation safety roles.

Supporting these I.T.s were Engineering Assistants: A. L. Case, R. P. Cumby, and K. C. Knight. Within this large group were Maintenance Supervisors: J. D. Blanton, R. L. McKinney, J. Miniard, E. W. Sparks, and G. G. Underwood who had shops specializing in different phases of I&C support to ORNL. Another engineering group was Radiation Detection under R. K. Abele. Instrument Checker W. C. Clowers worked with Engineering Assistants C. E. Fowler and V. C. Miller. Product Design and Fabrication under G. A. Holt had Instrument Technicians: R. H. Anderson, Jack Campbell, C. D. Gibson, J. M. Groover, S. U. Hamric, C. J. Keathley, J. H. Knox, C. G. Ruffner and S. E. Worley. **Engineering Assistants:** G. A. Hamilton, S. K. Inman, W. W. King, R. A. Maples, A. A. Smith and C. E. Stevenson. **Maintenance Supervisors** Gerald Hamby and C. H. Tucker guided the fabrication effort. Process Instruments Maintenance under P. W. Hill ably manned by Instrument Technicians: C. T. Alexander, C. G. Allen, B. M. Anderson, W. N. Baird, H. A. Barnett, L. E. Basler, R. R. Bentz, R. P. Boissineau, B. L. Carpenter, T. E. Chambers, R. Chambers, R. E. Cox, B. C. Davis, M. A. Denkins, K.W. Dresher, W. H. Hicks, R. K. Hopson, R. E. Hutchens, J. D. Keller, J. L. Lane, T. M. Lewis, B. L. Love, J. W. McNeillie, J. C. Mee, R. L. Miller, W. R. Mosley, J. A. Ramsey, J. D. Richardson, K. J. Scott, R. F. Spille, R. L. Stansberry, D. L. Thomas, A. C. Tinley, C. H. Vineyard and D. C. White. At that time the total number of hourly was 112, weekly 59 and monthly 154 in the Division.

As the reader can observe from these organization lists, a massive effort was focused on the development and manufacture of unique instruments, controls, safety systems, and support of other Laboratory production and research equipment. Bill Eads remembers large craft groups and shops being captive to specific divisions or facilities such as HFIR, ORR and M&C. C.

T. Carney was foreman at the High Flux Isotope Reactor. Jim Farmer was foreman at the Oak Ridge Research Reactor, and Ray Tucker was foreman in the Metals and Ceramics Division activities. A 1985 organization chart shows R. H. Brown as foreman in the 6000 area. This Physics Division work was in support of the 25-MEV Tandem Accelerator and ORIC Accelerator at the Holifield Heavy Ion Research Facility, and the Oak Ridge Electron Linear Accelerator.

Oral Histories

Oral histories of this era are a primary source of our story. Thus they may leave out important segments of the work. Typical of the work during this period of time, the writing is presented as an intermingling of the recall of the recall of supervisors and I.T.s as the work was also the joint responsibility of supervisors and I.T.s.

George Kwiecien Recalls

George Kwiecien recalls: "THINKING BACK, interesting aspects of my job with the I&C Division, from 1965 to 1987. He retired from the US Marine Corp after 20 years of service, serving in Korea in the fifties.

"When I joined I&C in 1965 I was almost immediately assigned to maintain the Environmental Radiation Monitoring Instruments. There were three stages of this equipment depending on their proximity to the Lab. LAM's local air monitoring; PAM's perimeter air monitoring and RAM's remote air monitoring. The Perimeter Air Monitors were located at the old gates, except for one. Their locations were: Blair Road near K-25, the old bridge on Hwy. 58, Bethel Valley gate, Y-12 gate in O.R., on Oak Ridge Turnpike in the west end of town, across the lake from the EGCR, and I believe there was one on the Clinch River near Melton Hill Dam., I believe there were 22 of these.

ORNL 92208

Perimeter Air Monitor – air sampler monitors located at various locations around the ORNL-K-25 and Y-12 Plants and also at government installations throughout the Oak Ridge area

The PAM's were located outside the fence but pretty close behind - K25, at White Oak Lake, down Bethel Valley Road near the Quarry. The Remote Air Monitors were also at the TVA dams or other government owned land. These locations were: Cherokee, Dale Hollow, Douglas, Fort Loudon, Great Falls, Norris, and Watts Bar. There were 8 of these. The information from the LAM's and PAM's communicated by land lines into recorders located in Building 4500S. These instruments were calibrated weekly using a known source. The remotes were brought in each week by a HP surveyor and replaced by a freshly calibrated one. They had their own recorder attached to them. They could be located as far as 30 miles away. The Lab also had to furnish foliage clippings. These clippings would be burned and then read for radiation level. Either I or a Health Physics person would pick them up. We picked a box of grass one day and by the time we returned to the Lab the phone was ringing, the family's 5 year old son had left his turtle in the box of grass and was afraid his pet would get baked. He was all smiles when it was returned.

Another interesting part of my job was my work with a device called the Flounder. It consisted of 12 GM tubes in a parallel configuration enclosed in a watertight plastic housing. The Lab owned a boat at the time called the Blue Goose. The Blue Goose would take the Flounder down the Clinch River each summer, dropping it to the bottom at predetermined spots, using the same spots each year. The first drop was right below Melton Hill Dam and they continued almost all the way to Chattanooga. The Flounder was calibrated prior to each trip. Two criteria had to be met for this calibration, it had to be done in an area with a very low radiation background and this level had to remain constant from year to year. We did this calibrating at a facility called Katy's Kitchen, just off Bethel Valley Road.

The "Kitchen" was a Lab set up in a cave, it had some significant use during the Manhattan Project. Katy's Kitchen was named after the wife of an early official, but its real use was never as a kitchen. It was a good place for near zero radiation background calibrations. (John Blanton adds that the Blue Goose went further than Watts Bar Dam. They monitored as far as Kentucky Lake every few years.) I would normally do this calibrating on a Saturday when the Kitchen was quiet.

In about 1980 I was assigned to do instrument work at the Tower Shielding Reactor. The work was primarily on the reactor safety controls and the fence monitors, these were located at various spots on the Tower Shielding perimeter fence. I also was responsible for TV cameras that were located at both the 100 and 300 foot levels of the towers. There was a ladder and an elevator that took us to these sites. The 300 foot level was quite a place to eat your lunch on a clear day. On a couple occasions I was asked to accompany families to their family cemeteries inside the Tower fence. Security was not as critical an issue as now and so I could take them to their site without a lot of hassle. It was very interesting to hear them talk of life in those hills back in the 30s.

I also spent quite a lot of time calibrating Dosimeters. These were the instruments that read the charge on the old pencil type radiation detectors. These pencil meters were simply a very high quality capacitor that would lose some of its charge when exposed to a radiation field. The pencil meters were charged to 300 volts, and the Dosimeter would read out how much of that charge was left and translate that to how much radiation the wearer was exposed to. Each person carried two meters, they were turned in at the gate at the end of the day and a new set picked up each morning. The meters were read each night. any high reading noted, then recharged and reissued."

Don Miller Remembers

Don Miller, who was hired in 1963, remembers the close knit relationship between Division Sections and their assigned craft personnel (Instrument Technicians, Foremen, and Engineering Assistants). Each section had Instrument Technicians permanently assigned according to the major projects under way and the general ORNL work requiring instrument support. Each shop or research laboratory had lead Engineers who, in effect, coached the craft supervisors and I. T.'s to achieve the desired technical and quality result on a given project. In the midst of the expansion of I&C staff and work variety, opportunities for promotion from the maintenance department to the engineering groups was frequent. Lou Thacker recalls transferring W.R. (Bill) Miller from Maintenance to Analog Circuit Development. Bill had demonstrated a particular facility to work from a sketch on an envelope to creation of a working analog circuit in a very short time.

When the Atomic Energy Commission made the decision to shut down the Oak Ridge Graphite Reactor in 1963, the 20th anniversary of its startup, shutdown planning included a significant "face lift". Instrument Technicians led the effort with several weeks refurbishing nameplates on controls, removing obsolete or unneeded equipment, rerouting thermocouple cables on the reactor face, and organizing maintenance manuals and drawings. Due to a control failure in one of the experiments associated with the reactor, water seeped through the top cover of the reactor and soaked the graphite. Operations and maintenance staff were faced with the dilemma of removing the water with 30 days remaining before the official shut down. The water soaked graphite prevented the reactor start up. Fully withdrawing all control rods was ineffective in producing a chain reaction. Subsequent discussions resulted in a plan to build a log scale power indicator simulator. If the reactor wouldn't start, we would at least have the power indicator demonstrate the effect of shutdown for the visitors. Karl West and Don Miller set about building a circuit that could simulate the Log Scale signal and associated reduction of power level at a rate similar to that achieved by inserting all control rods. The simulator design and construction was authorized just days before the official ceremony at which AEC officials were to be present. At the same time, dozens of industrial space heaters, called salamanders, were acquired to provide heated inlet air for the reactor cooling system in an effort to remove sufficient water to permit a normal restart. The simulator was completed and tested before the big day, but fortunately the drying process was successful and the dignitaries witnessed a real power reduction at the ceremony.

Don Miller placed his initials and the date on one of the strip chart recorder charts as he turned off the power to the Instrument for the last time. Don Miller remembers being "assigned to the Reactor Systems Design Group, under the leadership of John Anderson, during the height of the second generation reactor controls design. "It was stimulating work requiring design, prototype fabrication and testing of each nuclear channel module. A worldwide search was conducted to identify electronic components and hardware which offered the best stability and life expectancy. My job, as a loaned hourly from maintenance, was primarily to build the modules and connect suitable cables permitting installation in an environmental chamber for stress testing. In addition to familiar functions such as power supplies, special modules were designed and tested. The High Flux Isotope Reactor required safety systems to react faster than anything ever built before. One such module was the Fast Trip Comparator. Being solid state and pushing the limits of available transistors, required us to buy batches of transistors to bench test for beta and leakage characteristics. Often only ten or twenty percent of the batch was selected for use in operational modules. " Typical of HFIR instruments is the Electrometer shown in the picture below:

ORNL 98082-69

This Electrometer is typical of the instruments used in the HFIR reactor. It was fabricated in Clint Courtney's shop.

Beginning in the late 1960's and continuing through the 1980's craft connection with the engineering development groups began a migration to a less formal connection with I&C section and group managers. This trend was driven by a reduction of funding for major programs and growth of numerous smaller and technologically unrelated research works. Although individual Instrument Technicians continued to be assigned to some Laboratory development programs, many returned to a central shop environment.

This period found new and smaller development activities in I&C that could best use the expanded maintenance activity as needed rather than full time. It was in this environment that Craft Supervisors took on a more central role as arbitrator and coach to respond to manpower requirements from research groups and at the same time respond to mature ORNL operations that required regular Instrument Technician support to assure product quality and timely delivery. In this period the I&C Division also responded to changing funding and national energy focus by forming development groups that specialized in application of new commercial technology to Laboratory programs. Again the need for highly specialized Instrument Technicians became apparent. At this point the apprentice program had become too expensive and narrow in scope of technology and was terminated. The labor market was now rich with highly trained and experienced craftsmen who desired to associate with a national laboratory. Burroughs, IBM, AT&T, US military, other national science laboratories, and many more were sources of these new employees. Many new Instrument Technicians, as they were hired into the journeyman classification, went to work immediately on I&C and non-I&C customer needs.

Bob Vines Recalls

Bob Vines recalls being assigned to Charlie Allen's shop after being hired in 1979. Bob had come from Denver, Colorado where he worked for Martin Marietta Aerospace. During his time on the bench he and A. J. Beal were bench partners. For the uninitiated that meant they had adjoining personal work benches. Shortly after coming to work Bob was transferred to the Steam Plant to replace an IT who had just had a heart attack. Later he returned to 3500 for a time and then was transferred to the TRU building in ORNL's reactor valley. A new and somewhat eclectic development opportunity came to the division through the customer contact of Gerald Sullivan. The project contained both a short delivery deadline and the need to find low cost effective use of technology to control airport runway lighting. This was a task that Bob and Gerald teamed up on. With Bob's leadership, the hourly in his shop were empowered to find existing technology to mix and match to quickly develop a working demonstration device. An example of one creative solution was [H.L.] Haga's idea to use wireless automobile door lock control for battery disconnect. Through the years of the 80's Bob demonstrated a strong sense of practical solutions to industrial control challenges on numerous tasks. An Air Force job required the creation of a fire alarm system for large computer installation. Later in the mid 1990's he was promoted to staff engineer and assigned to Jim McEvers design group. His first assignment was the ORNL "Tank Farm".

Charlie Mossman Remembers

Charlie Mossman remembers that all written communications was by mechanical typewriter, six carbon copies or a stencil for the Ditto machine (hand cranked). Organization charts and other graphics were done by the Division draftsmen. Xerox introduced the first model 914 (copier) in 1949 and he doubts if I&C got one before 1965. Thus detailed written reports for work in that era are non-existent now. In this period Bill Ladniak and Charlie Harrell had offices on the front of the new one-story 3500 building. They each had a secretary and in addition there was one typist, Nell Newman, who did all the typing for the engineers. In addition, Nell could be a little feisty about your handwriting, spelling or grammar. A popular folk tale with the hourly I&C staff was speculation about the primary task, which occupied Charlie Harrill's time. Charlie's work style often included working issues in his office and developing plans and reports at his desk, resulting in the impression (to some passers by) that he had sufficient free time to count bricks in the building across the street. The second story of building 3500 was added in the late 1950's.

Mossman says the general strategy in this time frame was to form a team of I.T.s and Engineers to perform design and construction of new projects. Examples include controls, safety systems and research data monitoring and recording for the Homogenous Reactor (HRE), Oak Ridge Research Reactor (ORR), building 3019 Pilot Plant and OREX[1] (Y-12).

In addition to the Aircraft Reactor Experiment (ARE), the Bulk Shielding Reactor (BSR), and the Tower Shielding Facility (TSF) built as part of its Aircraft Nuclear Project for the Air Force, the Laboratory had three other major reactor designs in progress during the mid-1950s: its own new research reactor with a high neutron flux; a portable package reactor for the Army; and the Aqueous Homogeneous Reactor, which was unique because it combined fuel, moderator, and coolant in a single solution.

Maintenance Groups were also assigned to major facilities, and are enumerated as follows: 1500, 2026, 2519, 3010, 3019, 3042, 3047, 4500-S, 5500, 6000, 7500, 7900, 7920 and Tower Shielding Facility.

1. Metals and Ceramics in building 4500
2. Reactor Systems; Buildings 3001, 3042, 7503, 7900
3. Radiation Devices and Monitoring; Building 3026
4. Radio Security and Teletype; Building 3517
5. Physics; Building 6000

A major maintenance activity began to emerge as development of new programs required semi-permanent installation of monitoring and control equipment for operations. Bernie Lieberman remembers the development by I&C of a standard instrument panel design with process flow sheet graphics on the face. These panels required the skills of a relatively new breed of Instrument Technician who had both mechanical, electrical, electronic and graphic fabrication expertise. David White was one of these unique individuals who came from the petroleum industry to I&C and was immediately set to work on fabrication.

Richard Mathis Remembers

Richard Mathis remembers, "I&C transportation was very scarce. Most I&C shops had a unique aluminum two-wheeled heavy-duty cart that was designed and built in plant to move instruments between customers' locations and the shops. Electronic stores was located behind building 3500 for many years and was convenient for those working in 3500. Most shops had their own stock of commonly used parts. In the early years, the plant stores system was open in that customers could go back in the bins and shelves and get what they needed, fill out the appropriate IBM card and go on their way. At this time, clerks helped people find what they were looking for, stocked shelves, and processed the IBM cards. At some point in time, they went to a closed stores system where no one other than a stores clerk could go behind the counter without an escort and the clerk had to pull all items from the shelves and bins and process the check out documents in addition to their other duties. This did reduce the problem of people not accounting for the items they removed from stores, but many man hours of technician time was wasted waiting in line for items needed in the repair and fabrication process.

In the 1950's and before, instrument circuits were mostly vacuum tube and analog based. Readout devices were also analog with a few electro-mechanical printers. The transistor was new and expensive and didn't make much of an impact in the instrumentation field until the 1960's. In the 1960's, I&C started training its technicians in transistor circuitry since new instrumentation and I&C designs were now taking advantage of a wide variety of multipurpose transistors and solid state circuitry that were then affordable." Richard remembers the continuous application of new technology, such as analog readout devices. Don Miller remembers being invited to attend an RCA Institute Transistor course that was brought to 3500 for several days of instruction. The course included lecture and bread boarding of basic amplifier circuits. Remember that ones and zeros came later as integrated circuits came into general use.

Karl West Remembers

Karl West remembers a classic situation for I&C Technicians, who were often at the forefront of technological development in nuclear reactor and hot cell startup.

I.T. work associated with nuclear reactors required workers who had a special blend of knowledge about physics, electronics, integrated systems, and human relations skills. Each reactor required an I.T. on each shift who worked with a team consisting of health physics technologists, reactor operators, shift supervisors and experimenters. Ultimately, the reactor was there for the experimenter. Often I.T.s were required to multiplex their time between reactor systems support and experiment maintenance or fabrication of research instruments. During shut-downs of the reactor, everyone changed their primary focus to work that could only be done during shutdown. I.T.s usually focused on reactor controls, reactor and experiment monitoring equipment, and replacement of control modules. The work was dominated by strict paper trails and quality control procedures. Often the reactor down time seemed to be a 24-hour beehive of activity. The ORR story above is typical of the early years of reactor operations.

With the help of experienced chemical engineers brought to the Laboratory after its acquisition of the Y-12 laboratories, the Laboratory proposed to address these design challenges. George Felbeck, Union Carbide manager, encouraged their efforts. Rather than await theoretical solutions, Laboratory staff attacked the problems empirically by building a small, cheap experimental homogeneous reactor model. Engineering and design studies began in the Reactor Experimental Engineering Division under Charles Winters, and in 1951 the effort formally became a project under John Swartout and Samuel Beall.

This was the Laboratory's first cross-divisional program. Swartout provided program direction to groups assigned in the Chemistry, Chemical Technology, Instrumentation and Controls, Metallurgy, and Engineering divisions, while Samuel Beall led construction and operations. The OREX process, in which an **OR**ganic solution of lithium was **EX**changed with a solution of lithium in mercury or an amalgam, never advanced further than the pilot plant stage. The OREX pilot plant in 4501 was built in 1952 and subsequently dismantled between 1957 and 1959. This was the largest project at ORNL to that date that involved an extensive I&C design effort. The design criteria led to a breakthrough initial use of modular panels with process specific graphics associated with control valves and pressure indicators.

I&Cs maintenance staff was characterized by an unusually broad skill set and range of daily tasks supported. The Radio Shop is a good example of this breadth. Almost from the beginning, the Radio Shop provided maintenance and calibration of specialized radio frequency devices, sound systems, security equipment and scientific meeting support. One example of this support was the llth annual meeting of the American Nuclear Society in Gatlinburg during the week of June 22, 1965. A major focus of this meeting was Reactors in Urban Areas as reported in the Knoxville News Sentinel. The Atomic Energy Commission has begun a study aimed at providing assurance that large nuclear power reactors can be located safely near metropolitan areas.

New Management Structure Needed

By the mid 1970's it became clear that I&C management would have to reorganize for a separate management structure totally dedicated to hiring, training, work assignment and union conflict management. In the rush to respond to early program needs and later to development group specialization, management had accumulated a large backlog of union grievances. It was during the 1950 to 1970 time frame that the Atomic Trades and Labor Council, a bargaining unit for all trades in Oak Ridge and a consortium of 16 A.F. of L. craft unions, became effective and active in pressing for better benefits for its constituents. The ultimate union structure was a result of an election held by the Department of Labor at the X-10 site. A byproduct of that effort was the empowerment of each craft union group to become active in monitoring the detailed day-to-day activities of engineering groups and Instrument Technicians in shops. For the uninitiated, Instrument Technicians were the best-educated and most creative group among the various crafts. They were required to understand a broad range of technological, scientific and fabrication skills to be effective. Although complex to describe, to the uninitiated, a simple example about to be made.

A hypothetical reactor control room supervisor has requested that I&C replace a metal panel, which currently contains a 1950 vintage temperature recorder capable of continuously recording two temperatures from the reactor vessel. The goal is to install the current technology twenty channel temperature recorder, and include a pneumatic transducer and heavy duty A/C power relay. In order to produce the most professional look and best quality control of the new panel, Instrument Technicians are apt to punch holes in the panel, mount equipment and test the functions of all the devices before delivery to the reactor for installation. Thus they have assured 100% likelihood of no rework because of discovered flaws after installation. The catch, union craft wise, is that the pipe fitter at the reactor may choose to take issue with the IT performing "his work" by installing and or testing the pneumatic transducer (it has plumbing). Likewise the electrician assigned to the reactor might well take issue with the IT installing and or testing the power relay. This one, relatively simple job, has now produced two grievances. In union parlance, two grievances require two craft supervisors and two union representatives to meet and debate the perceived crossover of craft responsibilities, with a goal of reducing future such discussions.

During this time frame, the grievance list had grown from a few per year to several hundred unresolved by 1979. To this writer, the two primary reasons for the establishment of the Maintenance Management Department were programmatic and union/management challenges. It should be made clear that this evolution was a natural outcome of changing programs and regulatory expectations. This new arrangement permitted engineering group leaders to give primary focus to technical issues at a time when each national laboratory was finding itself competing for federal program dollars. Also, the fraction of all IT work that was repetitive in nature had grown significantly over the years. Numerous large processes such as nuclear reactors and hot cell operations were permanently part of the daily routine.

It is not surprising that the organizational structure, initially developed, for the new department was chosen to split the numerous different shops and individual assignments into two major general sections with similar technological focus. This offered an efficient use of intermediate level supervisors and yet provided a manageable technology knowledge requirement for maintenance managers. "Process and Environmental Instrumentation", and "Communications and Data Instrumentation" were chosen for the primary halves of the new department. This arrangement was workable and sustained until the year 2000 as the primary services offered. In simple terms, one group relied on its familiarity with the earliest technical challenges at ORNL and the evolving technological advancements; The measurement and documentation of radiation issues had a regulatory and practical operational emphasis. Communications and Data Instrumentation was strongly based in semiconductor technology, and the transmission, recording and storage of scientific data. The hardware of this activity was often computer based and required interconnection with Laboratory networks, data storage devices and individual work stations.

Selection of a suitable individual to lead this new Department was relatively easy after listing all requirements. Foster a team spirit with the newly empowered maintenance team, and deal effectively with the grievance backlog. Paul Hill was selected from within the organization because of his easy going nature and natural leadership qualities. In a relatively short time, Paul reduced the grievance backlog and fostered a harmonious craft and management relationship. Meanwhile the I&C Engineering groups and major projects were adopting new technology and fabrication techniques. The Oak Ridge Research Reactor upgrades, Molten Salt Reactor's use of computer data acquisition systems and the early waste management instruments were switching from vacuum tube technology to solid state devices. These projects were increasingly adopting printed circuit instrument construction.

Except for a few Technicians hired from the computer industry, few were familiar with solid state circuit devices. Technology current staff such as Ray Cinnamon, Bill Miller, Jim Jansen, Bill Bryan, Clint Miller, Bud Cooper, and several others were assigned to teach short courses from time to time. This internal self training sustained the huge technological evolution of equipment being designed and supported by I&C. Some technology required the importation of training from the electronics industry for hourly and weekly staff. RCA Institutes, for example, was hired to teach a 40-hour course in solid state devices to most of the I.T. classification.

Don Miller Wore Two Hats

During this volatile year for I&C, Don Miller was wearing two hats. He had built a solid work base in Energy Division beginning with the Annual Cycle Energy System research on the University of Tennessee's agricultural farm. While initiating the modernization of tools and training in maintenance he was concluding several energy conservation jobs. The Fort Stewart, Georgia Wood Burning Boiler Plant had recently installed a wood burning boiler for barracks heat and central laundry hot water. This was one of a dozen projects resulting from the energy crisis of the 70's. Energy Division had contracted to validate these projects by retrofitting instrumentation to measure total energy consumed and estimate the oil equivalent that would have been used in this case. Other ECIP projects which Miller contributed to were Fort Huachuca's central HVAC retrofit and Fort Bragg's computer controlled barracks temperature retrofit.

Returning to the I&C Maintenance Department's challenges, ORNL was facing strong competition from other national laboratories for research and development contracts. A major foundation structure needed to assure credible science work was the electronic calibration infrastructure and traceability to national standards. In earlier years individual scientists were responsible for this traceability. As projects grew in size and number the central standards laboratory became more important to the overall mission of the Lab and to the increasing capability of DOE to audit our procedures and processes. This dynamic required a significant budgetary and organizational commitment to strengthening I&C's field shop capability.

Although, initially, Miller's focus was on equipment acquisition and basic calibration training, it soon became obvious that a more comprehensive systems upgrade was needed. The original goal of MAINS was to respond to the need for automation of the data collected and shared with other divisions. The new urgency was both traceability of standards and trending of information to assure system wide transparency and detailed trend analysis capability. From this realization came a focus on nuclear and safety measurements. It was necessary to know the actual error of a measurement and the rate of change over time. Through the use of internal staff, summer graduate school participants and volunteer engineering help from our I&C sections Maintenance, Accountability, Jobs and Inventory Control (MAJIC) Program became a reality. Later in a MAJIC technical manual it was said that the system be maintained in a way that assures the collection and reporting of data needed to comply with DOE policies and procedures (regulations).

A New Era Emerges

Ray Cinnamon was an hourly employee who advanced quickly to monthly pay grades and was a frequent instructor for Instrument Technicians teaching transistor circuits and later digital circuit logic. Shown below is a picture of Ray Cinnamon teaching a microprocessor class.

Members of a Microprocessor class are taught by Ray Cinnamon.

Analog readout devices were being replaced with digital readouts that were first various styles of glow discharge tubes, then came the backlit numerical panels, to the seven segment displays, to the LED based displays, to the LCD displays. I&C had its own printed circuit shop and fabricated thousands of circuit boards for special use and I&C designed instruments. Robert Maples worked in this shop for many years. At first it was single sided boards, then double sided boards. When commercial manufacturers introduced multi-layer boards, the in house board fabrication began a gradual decline in popularity with designers. Richard continues: "My first assignment was in a process control shop supervised by Jim Day in building 3500. This shop maintained electronic and pneumatic control instruments and readouts. In this shop was a recorder repair area which specialized in Brown and Leeds & Northrup recorders and controllers. Howard Frazier and Jim Knox worked in the recorder shop. I first worked with Charlie Hicks. Later I worked with Jack Richardson after Charlie left the company and went to work for ARAMCO in Saudi Arabia. Jack's specialty was pneumatic instruments of which Foxboro and Taylor were popular name brands."

Introduction To Section 2

By Ray Adams

This section contains a summary of the bulk of the I&C effort during the years 1954 to 1990. During much of this time, the Assistant or Associate directors (the title changed over the years) of ORNL for Nuclear and Engineering technologies were Walter Jordan (himself an original Instrument inventor [the A-1 amplifier]) and later, Don Trauger. Both of these men were strong and personable leaders. They are pictured below, Don Trauger (in the picture on the right) is shown with Cas Borkowski, left, the head of the I&C Division during much of this period..

Walt Jordan, ORNL Associate Director and friend of the I&C Division

Don Trauger, ORNL Associate Director during much of the I&C Division existance (shown with I&C Director C.J. Borkowsxi – on the left)

In the early part of this period, the I&C Division was composed of only two departments, The Reactor Controls Department and the Instrument Department. Each of the departments included instrument technicians, who were closely allied with one of the engineering groups. The engineering groups relied upon the instrument technicians to construct and maintain their "special" kind of instruments and control systems and there developed thereby a sub-specialty in that sort of instrumentation and controls apparatus.

Thus, there were technician specialists in electronic measurement systems, digital technology, reactor safety systems, etc.

The Reactor Controls Department was newly formed, when the I&C Division was founded. It was headed for many years by E.P. Epler. Some of the work of that department is covered in Section 3 of this book. However, the Instrument Department was brought over from the Engineering and Maintenance Division and had existed from the beginnings of Clinton Laboratory days (see photograph [VertClinton Lab.tif], page xxx). Much of the work of the Instrument Department is covered in this section.

Some of the work of the Instrument Technicians is covered in this section. They considered themselves as the creators and caregivers of the instruments - frequently voicing their work as "from the cradle to the grave" for their charges. Their work and the content of their energy left them time for such things as first-aid and sports competitions. There is a rumor that at least one Instrument Technician was hired for his ability to play softball and teams comprised of I&C technicians won more than their share of the competitions held by the recreation department. The Instrument Technicians were always a part of the hourly bargaining unit that included electricians. However from time to time, they made an attempt to form a separate bargaining unit, but never could quite "pull it off" as the bargaining unit did not recognize that sort of structure in their traditional organization. No small part of the position of the I&C Division as "Beyond the Edge of Ttechnology" was due to the skill of the Instrument Technicians.

Chapter 6

Instrumentation and Controls Engineering – Particularly Electronics, 1954 and Onward

Ray Adams

For several years, after 1953, when the I&C Division was formed, the Instrument Department, headed by C. S. Harrill, remained pretty much as it was under the P&E Division. There were two major subdivisions one, a Service subdivision in which the maintenance and construction were managed, and a Research subdivision, under which the engineering was conducted. Leon Reynolds initially headed the research subdivision. There were several pioneers in the ORNL Electronics design business, who pre-dated the formation of the I&C Division.

Pioneers in Electronic designs
Floyd Glass, Hugh Wilson, Walter Jordan, Ed Fairstein

The I&C Organization was essentially the following:

Electronics Section – Frank Manning

Early (1954) Members of Manning's Section

Engineers
    Wil Adams, Byron Behr, Odell Eason, Tom Emmer, Floyd Glass, Gerry James, Ken Kline, Layton Meeks, Bob Sidnam, D.D. Walker, and Hugh Wilson.

Technicians
John Francis, Ralph Gitgood

Electronics Section - 1966 Org. Chart

Frank Manning plus Group Leaders:
Harry Todd – Electronic Services for Neutron Physics
Dave Knowles – Monitoring Systems Development
Jim Lovvorn – Special Electronics Services
Hugh Wilson – Circuit Development
George Holt – Product Design & Fabrication
Jim Woody – Computers & Pulse Techniques
Woody retired around 1976
Ed Madden became group leader.

Other Engineering Sections

Based upon 1966 Org. Chart
Process Control and Instrumentation (X-10 Mossman)
Reactor Projects (Y-12 9201-3 Metz)
Reactor Projects (Y-12 9204-1 Moore)
Graphics - Mark Bowell
Mechanical Development - Stripling
Allin, Sliski
Radiation Detection – Roland K. Abele
H.V. Projects - Jim Johnson

Various Projects in the Electronics Section

Ed Madden's recollections

In 1961 the Oak Ridge Isochronous Cyclotron (ORIC) was under construction, to be housed in building 6000. It was to be a variable-energy accelerator using a fixed radio-frequency excitation and an azimuthally-varying magnetic field to accelerate a variety of ions from protons to highly charged krypton atoms over a wide range of energies up to 145 Mev (Million electron volts). It was designed to produce one milliampere of 75 mev protons. It employed a radial increasing magnetic field to maintain isochronism and azimuthal variations in the field, to provide radial and axial focusing. It employed a 208 ton water-cooled magnet with 76 inch diameter pole tips divided into three slightly spiraled sectors. Physicists at the lab had designed the ORIC and they provided oversight in the construction and final operation. The Oak Ridge Isochronous Cyclotron Facility was built with 19 magnet power supplies which were rectifier units controlled by a combination of saturable reactors and transistors. I recall that Bill White from I&C was working with the power supply units. Four sets of coils were employed, including the main coils, a set of valley coils polarized oppositely to the main coils to enhance azimuth variation, a set of harmonic coils to control the orbit center, and circular trimming coils to obtain the isochronous field for various ions. The large building included office and laboratory space, provided two large separate experimental areas to which the beam could be directed by a magnetic switching system,

The Instrumentation and Controls (I&C) Division provided manpower for various instrumentation tasks in the final stages of completion of this facility. I worked with Jack Russell in checking wiring throughout the building. All the wiring had already been installed by the contractors, but it was essential to determine that it was all correct before attempting to apply power to the cyclotron. We prepared daily lists of point to point wiring to check, and then worked with the electricians to check each and every wire to make sure that the miles of wire in that building were properly installed. The actual work was not all that challenging. Racks and racks of construction drawings had to be explored and many lists of wires generated. Good crews of people worked in the building, everyone cooperated, and it was a very enjoyable experience, which lasted less than a year, probably about 6 months. I met a number of people during this time that became friendly contacts throughout the years that I worked at ORNL. At the end of this project I was moved to the 3500 building to work under the group leader Hugh Wilson in the Circuit Development Group, which was part of a Section under Frank Manning.

In 1961, the Electronics Section of the I&C Division was involved in development of nuclear spectroscopy instrumentation, fast-neutron survey meters, Geiger Mueller survey meters, silicon surface barrier detectors, alpha scintillators, whole body counters, containment instrumentation. The 5Mev and 3Mev VanDeGraff accelerator instrumentation was directed by Gene Banta's group.

In 1962, the Circuit Development Group under Hugh Wilson consisted of Nat Hill / Bob Scroggs /Jim Madison / TA Love / Floyd Glass / Jim Lovvorn / Jim Todd / Joe DeLorenzo, and myself. In 1962, there was work on; transistor pulse amplifiers, 150Mev cyclotron experiments, magnet regulators for ORIC, Photo-multiplier tubes, Scintillation systems, surface barrier detectors, Radiation monitoring, Containment instrumentation, alpha and beta-gamma monitors, and civil defense instruments. That year, The Oak Ridge Institute of Nuclear studies gave a two-week course in Radiation Physics and Fundamentals to many employees, which I attended. Precise measurements of radiation activity were required to monitor and preserve a safe environment and to ensure nuclear power reactors to be designed were safer, cheaper, and more functional.

Howard Burger (of Bob Moore's I&C Reactor Projects Group) brought a problem to Hugh Wilson's group that needed solving. He had invested a great deal of time and money in a scanning system for the Molten Salt Reactor to monitor and display the signals from approximately 420 thermocouples. The reactor thermocouples were monitored through commutated contacts of a 100 point mercury jet switch, 100 contacts each, running at 1200 RPM, then routed through two double-pole double throw choppers operating synchronously and then through a DC amplifier. They were then displayed on a 17 inch oscilloscope. An alarm detector fired when the temperature difference between the sampled signals and a reference signal exceeded a preset value. The thermocouples were scanned at a rate of 2000 points per second. A number of the foot tall, 8 inch diameter, mercury switches had been purchased before it was discovered that the switch was so noisy that the noisy signal overpowered the input signals from the thermocouples on the molten salt-reactor experiment.

After a number of days on the bench, I tried different orientations of the switch. I ran the switch sitting at about a 30-degree angle and the noise disappeared. The switch was very heavy and few would have attempted to run the switch sitting at such an angle. I totally disassembled the switch, and careful examination of the construction led me to the conclusion that the addition of a stainless steel grounding ring inside the switch would solve the noise problem. I had the machine shop fabricate a 6 inch grounding ring, modified the switch, and lo and behold the noise problem disappeared. All the mercury switches used in the reactor instrumentation were subsequently modified with a similar grounding ring. This illustrates how various groups within the I&C Division often worked to solve problems that would elude any one group.

Instrumentation and Controls Division work in FY 1963 included Semiconductor Radiation Detection research;

Transistors were used in logic circuits for the space program, neutron time-of-flight spectrometer instrumentation; time-to-pulse height conversion, count rate meters, remote control of accelerators, radiation fallout monitoring of exhaust gases, alpha proportional counters, Beta-gamma and gamma ray monitors, personal radiation monitors, parallel plate fission chambers, x-ray proportional counters, gamma ionization chamber, and radiation shielding. Other I&C involvement included; lithium-drifted silicon radiation detectors, x-ray fuel plate scanner, gamma ray scintillation detector, heat transfer systems, data logging, and reactor control rod instrumentation. Thermocouple calibration, test reactor instrumentation, ORR instrumentation, use of silicon controlled rectifiers, helium leak detection, and radiation dosimeters. Further items were; Health Physics research reactor instrumentation, molten salt reactor experiment instrumentation, reactor control and safety, ionization chambers, solid-state differential amplifiers, MSRE instrumentation, flow meters, strain gage measurements, thermocouple testing, and support for 3 Mev Van De Graff and Tandem Van De Graff

Computer and Pulse Techniques Group

Ed Madden's recall (cont'd)

1964 I&C involvement - In 1964, I was part of Jim Woody's Computer and Pulse Techniques Group, under Frank Manning as the section head. Billy Joe Moore, a class-mate of mine at University of Tennessee, had been working in Jim,s group when Billy Joe resigned to go to California and open up his own business. I was selected to replace Billy Joe in Jim Woody's group. In the fiscal year 1964, I designed, built, and put into operation an automatic film-badge reader and cardpunch system for the health physics department. An electromechanical reader head sequentially positioned a piece of developed film-badge film, from the ORNL film-badge dosimeter, to each of four absorber area positions in order. A densitometer measured each area of the film and provided a calibrated measurement, representative of the radiation exposure received by the film. Each reading was displayed on a digital voltmeter, and an IBM card was automatically punched with the collected measurement information. Every employee carried a plant badge that had a built in radiation dosimeter that indicated radiation exposure by the exposure to the film through a set of four different radiation absorber areas. Every employee exchanged his badge each month and each person's badge film was analyzed for beta and gamma radiation exposure. Analyzing the badge films of 3000 to 5000 employees each month was a large undertaking for the Health Physics Department at ORNL.

1964 I&C involvement - Time-of-flight instrumentation / gamma-ray anti-coincidence and pair spectrometers / multi crystal gamma ray spectrometer / plastic scintillators / single and multi-channel analyzers / time to pulse height conversion / time interval meters / punch tape data storage systems / solid state linear count-rate meters / radiation monitoring and control systems / fallout monitoring / monitoring radioactive exhaust gases / Laboratory radiation containment / alpha proportional counters / Beta-gamma radiation monitoring / Personal radiation pocket monitors / x-ray proportional counters / gamma ionization chambers / radiation shielding /

lithium drifted silicon radiation detectors / neutron time of flight / gamma ray scintillation

/x-ray scanner to scan reactor fuel plates - JW Reynolds / heat transfer systems / data logging systems / computer codes for data analysis / EGCR control rod tests / Thermocouple calibration computer programs / engineering test reactor instrumentation / ORR instrumentation / pilot plant instrumentation / high level radiation examination laboratory /radiochemical pilot plant / foam decontaminant studies / telemetering of liquid levels in waste storage tanks / measurement of hexavalent chromium for chemistry division / dynamic gas thermometry /vacuum ion and control / silicon controlled rectifier applications / testing and calibration of instruments in the field / inspection and calibration of thermocouples / Standards Laboratory / Helium Leak detection / airborne fluorides detection / photometer for emitted-light measurements / chemical dosimeters / EGCR instrumentation / Health Physics Research Reactor / ORR-GCR loop / low current transistor amplifiers / fast trip comparator circuits / electronically adjustable gamma-compensated ionization chamber / analysis of neutron fluctuation spectral density / low noise solid state differential amplifier / MSRE automatic rod controller / Mechanical drive for positioning a fission chamber at MSRE / process instrumentation / water flow meters / tests of strain gage pressure transducers / thermocouple materials and hermetic seals for sheathed thermocouples / 3 Mv Van De Graff accelerator / improved Klystron for 3 Mv / Tandem Van De Graff / Negative ion source development

1965 I&C involvement - The September 1, 1965 I&C Annual Report of an Automatic Film-Badge Reader and Card-Punch System by E. Madden. An electromechanical reader head sequentially positions a piece of developed film from the ORNL film badge dosimeter to each of the four absorber area positions. A densitometer measures each area of the film and provides a calibrated measurement, representative of the radiation exposure received by the film. Each reading is displayed on a digital voltmeter, and an IBM card is automatically punched.

1965 On Aug 25, 65, the ORNL High Flux Isotope Reactor achieved self-sustaining nuclear operation.

Editor's Note: The information following, includes mention of a large number of computerized systems. Some of these systems were employed by researchers for x-ray and Neutron diffraction research. Many of the computerized systems were the first use of computers for this type of work. Shown in the picture below is one such computerized system, which was the first fully computerized control and data-gathering application for x-ray diffraction. It was employed in the Chemistry work of Bill Busing and Henry Levy.

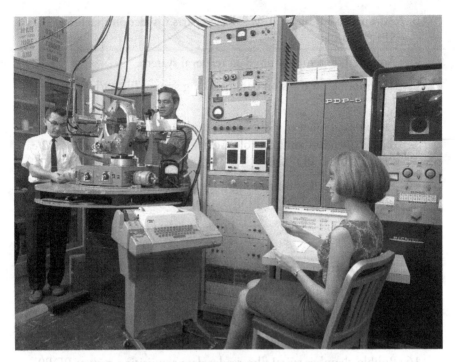

ORNL 13561-A

The first fully computerized x-ray diffraction apparatus. Shown are: L to R, Ron Roseberry (I&C Engineer), Ray Ellison (Chemistry), Sharron King (Math Panel). As mentioned below, numerous computerized systems for x-ray and Neutron diffraction studies, were employed throughout ORNL.

1970 I&C involvement - Data Systems work at ORNL by Ed Madden with computers involved:

1. Oak Ridge Electron Linear Accelerator (ORELA) data handling system 1 ea. PDP 10, 3 ea. PDP 810B, 5 ea. PDP15, 1 ea. SEL 840B, 1 ea. PDP 4,

2. Lunar Receiving Lab (NASA) Nuclear Data Acquisition and reduction system PDP 9

3. Oak Ridge Isochronous (ORIC) Cyclotron facility magnet power supply control and monitoring system MODCOMP III.

4. Astronomical Balloon Mount Gamma Ray Spectrometer data acquisition system for NASA.

5. Radio isotopic sand tracing data acquisition system for the US Army Corps of Engineers 2 ea. PDP8E.

6. Automatic double-arm sweep gage parts inspection machine SEL810B.

7. Aquatic Life temperature data retrieval system.

8. Thermonuclear division data handling system PDP8/E, PDP12, PDP11/45.

9. Eleven each nuclear spectrometer and x-ray/neutron diffractometer data acquisition and control systems 1 ea. PDP5, 6 ea. PDP8, 2 ea. PDP8/I, 2 ea. PDP8/E.

10. Four each nuclear time of flight data acquisition and control systems PDP 4, PDP 7, PDP 9, PDP 15/30.

11. Nuclear multi-analyzer stored program data handling system PDP8/E.

12. Mass Spectrometer data acquisition system IBM 1130.

13. Tower shielding data acquisition system PDP9.

14. High Voltage Laboratory data acquisition control CDC160A.

15. High flux isotope reactor x-ray fuel-plate scanning system PDP 8/I.

16. Bubble chamber spiral film reader data acquisition system PDP9, PDP8L

Woody retired in 1976 and Ed Madden became the group leader.

Radiation Detection Section

Marion Chiles

On February 18, 1957 I was hired by Roland (R. K.) Abele, Head of the Radiation Detection Section. At that time there were five technical people (referred to as Development Engineers) and two technicians in the section. Roland was one of the five as an Electronic Engineer. The others were Richard (R. E.) Zedler a Chemist and three Physicists, Vern (V. A.) McKay, Frank (F. E.) Gillespie and myself Marion (M. M.) Chiles. Ken Miller was an experienced technician in our laboratory and handled evaporations, vacuum pumps, soldering and brazing for all of us. Ken was approaching retirement so Taylor (W. T.) Clay was hired two months later to learn Ken's expertise in the lab to become Ken's replacement. Bee (B. R.) Parker assisted with testing, some fabrication and repair of detectors.

The primary responsibility of the Radiation Detection Section was to provide the proper radiation detectors to experimenters in other divisions at ORNL. Many of these chemists and physicists were conducting research with radioisotopes and would seek our advice about electronic equipment they should use including radiation detectors. Since there were not many commercial radiation detector manufacturers at that time, quite often we would have to design the detector to best perform their experiment. This was the challenging part of our job. Our section also designed reactor control chambers for the experimental reactors operating at ORNL. The Circuit Development Section and the Radiation Detection Section collaborated to provide many detector and electronic combinations for Health Physics monitoring and surveying. There were times we entered into national projects outside ORNL and even DOE contractors such as the National Aeronautics and Space Administration and branches of U. S. Military, to assist in their nuclear radiation detection endeavors.

The Development Engineers who had been in the group for a few years had become somewhat specialized in certain types of radiation detectors. Zedler was experienced in gas filled proportional counters, ionization chambers and GM tubes. McKay had become more specialized in scintillation detectors. Gillespie was heavily involved with providing fission chambers for experiments conducted by Neutron Physicist Division at ORNL accelerators.

Between 1960 and 1966 there were some changes and additions of personnel in the Radiation Detection Section. Charles Fowler was employed as a technician in the processing laboratory, (vacuum & gas filling systems, evaporations, soldering, brazing and many other tasks) along with Taylor Clay. During the time Clay was working as Technician in the laboratory, he was studying for a degree at the University of Tennessee. In 1962 he received a B. S. degree in Engineering Physics from U.T. and was promoted to technical status as a Development Engineer in the section. Also about that time Hugh R. Brashear (an Electronic Engineer graduate from Oklahoma State University) was employed as a member of the technical staff as an Engineer. Along about that time, Frank Gillespie accepted a job at Princeton University Laboratory and left the Section for approximately a year, then returned to his previous position in the Detection Section. In 1963 V. C. (Clint) Miller was hired by Roland Abele as a Technician in the detector processing laboratory with Charles Fowler. Roy L. Shipp (an employee of Lockheed Corp. in Marietta, GA) was on some cooperative program with ORNL and was assigned to Dave Knowles' group. He was involved with environmental radiation monitoring for the exhaust stacks and etc., which required close collaboration with the Detection Section. As a result of that affiliation, Roy was hired by ORNL and permanently assigned to the Detection Section.

The staff of the Radiation Detection Section remained much the same for a decade or so. Two members, Vern McKay and Frank Gillespie, died during that time, somewhere around the mid eighty's. About that time, the Radiation Detection Section became known as the Sensor Systems Development Group. During the latter part of the eighty's, Roland Abele was assigned another job in the division and Hugh Brashear became Head of the Sensor Systems Development Group. Hugh hired James Ramsey from the Instrument Maintenance Department as another Technician in the group. Not long after that, Hugh Brashear became Head of the Research Instruments Section and Martin (M. L.) Bauer became supervisor of the Sensor Systems Development Group. When Roy Shipp retired, Chet (A. C.) Morris transferred into the Sensor Group and about that time Stephanie McElhaney was hired into the Group and in 1993 Mark (M. A.) Buckner joined the Sensor Group. Chiles retired July 31,1994 and Fowler retired December 31, 1994. So far as I know the Sensor Group staff remained about the same till the I&C Division was dissolved into other divisions at ORNL.

Significant Projects of the Radiation Detection Section
2-18-57 through 7-31-94

My first assignment was to design a Reactor Control Chamber for the Tower Shielding Reactor. Of course being new in the group, I had close supervision by Abele. Those chambers worked well for the life of the reactor with replacement of the Teflon insulated coax connectors and refilling with nitrogen gas from time to time. Later I was given responsibility for designing some scintillation detectors because of my experience with photo-multiplier tubes at RCA in Lancaster, Pennsylvania before I came to ORNL.

In 1959 a request came from Ed Gupton in Health Physics Division for the detection section to design a scintillation detector to survey for alpha contamination with higher sensitivity than presently available. Several criteria were established as guide lines for designing such detector. We started with very simple, inexpensive, available components and designed for approximately 100 square centimeters of sensitive area. Methyl Methacrylate (Lucite) was the material selected to transmit the light (light-pipe) from the Zinc Sulfide [Zn(Ag)S] phosphor to the photomultiplier tube. After several thicknesses of Lucite were tested, 0.500" thickness was chosen. Then the design was completed using an inexpensive 2" diameter photomultiplier tube. Performance specifications were prepared for outside bidding and several hundred of these detectors were purchased for ORNL health physicist to use surveying for alpha contamination. As result of this simple design, these detectors were purchased from commercial manufacturers for less than $200.00 each, a basement bargain. Later in 1980's Stephanie McElhaney became an employee in the Radiation Detection Section and started developing a more rugged alpha scintillation detector. This became a group effort with technician James Ramsey, McElhaney, and Chiles participating. McElhaney took the lead in this and was successful in developing a detector with a light tight window that was more resistant to punctures than the conventional aluminum coated Mylar window. A patent was issued from the U.S. Patent Office for this design. This technology was transferred to commercial manufacturers and many were manufactured.

In the early 1960's there was a lot of Gamma Spectrometry being done primarily by Analytical Chemistry Division and others. This research promoted two efforts for the Radiation Detection Section to improve parameters involved with current equipment:

1- Improve performance quality of Na(Tl)I scintillation detectors; 2- Design lead (Pb) shields to reduce background radiation and more economically manufactured than shields now being used. Vern McKay and Marion Chiles were assigned to this task. Lead was selected to be the main outside material with no straight through cracks that would allow outside radiation to leak through as in stacked lead bricks commonly used at that time.

Effort #1-The first "Integral Line", 3" diameter , NaI(Tl) scintillation detector was designed so the crystal was optically coupled directly to the photomultiplier tube. This eliminated one optical interface and reduced the light scattering between the crystal and the photo-multiplier tube. This design established the state of the art performance for gamma ray spectrometry. Harshaw Chemical Corp. made this design one of their commercial line items. Later, Bicron Corp. did the same.

Effort #2- Discussions led to the suggestion of "graduated lining" inside the lead to reduce the (75 kev) x-rays produced by interactions from higher energy gamma photons with the lead atoms. This reduced the total count rate produced in the scintillation detector which improved performance of the electronic system, especially with more intense samples being measured. Chiles took the task of making absorption calculations using different metallic materials. Results of the calculations indicated that a liner of cadmium 0.093" thick would absorb most of the 75 kev x-rays from the lead and a liner of copper 0.010" thick inside the cadmium would absorb a high percentage of the 23 kev x-rays from the cadmium. These materials and thicknesses were selected to be used inside the lead shields and were used thereafter in many gamma ray shields for gamma spectrometry applications. For the shield, the concept was conceived of using 1" thick lead sheet and rolling it with radius of curvature according to the diameter desired. Three layers were stacked around each other arranged so no cracks overlapped. The final mechanical design was done by McKay and Ted Sliski in the Mechanical Department. These were manufactured commercially and many were used in Oak Ridge Labs. They were commonly referred to as the "Pickle Barrel Shields".

The Health Physics Division was funded to build a "Whole Body Counting Facility" to examine ORNL employees for Gamma-ray emitting radioisotopes. Some isotopes such as Potassium-40 are residual occurring in the body. But since the nuclear age came into prominence in the last sixty years, other isotopes present in the atmosphere have become somewhat residual in the human body of everyone. So the WBCF was built to research gamma radiation in people not working with radioactive materials as well as those working with radioactive materials. Employees were requested periodically to be tested in the facility to determine if there was any change in their body radioactivity. Vern McKay was assigned to design the gamma spectrometer and oversee the design of the shielded room and passage way in and out with door interlocks. The door interlocks were to prevent fresh air, which contains Radon and other background radioisotopes, from entering the counting room. The Gamma Radiation detector system consisted of two eight inches diameter by 4 inches thick sodium iodide Na(Tl)I scintillation crystals each optically coupled to three (three inch diameter) photo-multiplier tubes. One scintillation detector was positioned to scan the front of the body and the other one scanned the posterior of the body. The walls of the shielded room were constructed of stainless steel to shield against environmental background radiation and reduce backscatter radiation. The stainless steel selected for this facility had to have very low residual nuclear radiation, so radioactive qualitative analysis was required from each vendor submitting bids to supply the material for this job. This was required because at that time some stainless steel was being imported to USA from Europe that contained traces of some radioisotopes which would contribute to the background radiation inside the counting room. Later a high resolution Germanium detector replaced the scintillation detectors.

An Ion chamber with sensitivity of 400 cc air equivalent was needed for monitoring some process pipe lines at the new Molten Salt Reactor Experiment. Hugh Brashear was assigned by Roland Abele to design a one-inch diameter ionization chamber filled with increased pressure of Krypton gas to achieve the required sensitivity. The chamber was constructed of welded stainless steel housing and ceramic insulators for rugged operation. Specification was prepared for fabricating this chamber and commercial vendors manufactured many. Also a commercial electrometer was modified according to ORNL specifications and purchased to use with this ionization chamber at MSRE and many other monitoring locations at the Laboratory.

The Neutron Physics Division was carrying on several experiments to re-evaluate neutron cross section values for several fissile isotopes and also the number of neutrons emitted per fission. Frank Gillespie was assigned to design and develop fission chambers for their experiments. Several single plate chambers were designed as well as several multi-plate chambers. They were all parallel plate units with fissile material electroplated onto the substrate material, usually high purity aluminum. These chambers were 5 inches outside diameter with 3 inches diameter coating on inside parallel plates. In 1962 Frank Gillespie left the Laboratory to accept a job at Princeton Research Laboratory and Marion Chiles was assigned to work with the Neutron Physics Staff to provide their detectors. From the first designs they had realized that materials inside the neutron exposure beam besides the fissile material needed to be drastically reduced. Thin, rigid, high purity aluminum material was procured to make the inside plates. The end plates were reduced to only .005" thick. This made it very difficult to evacuate and gas fill the chamber after assembly. A system was designed to evacuate both the outside of the end plates along with the inside during the gas filling process and this worked fine. Several grams of fissile material could be used in these chambers to increase the statistical accuracy of their experimental data. Gillespie returned after a little over a year and resumed his work with the Neutron Physics Division. About this time, the Thermonuclear Division requested a similar chamber containing U^{238} and closer spacing between plates so the overall length was less. Chiles designed and developed such a chamber for them to monitor for theoretically possible high energy (1.5 mev) neutrons during their experiments.

The Thermonuclear Division also requested a high efficient neutron sensitive proportional chamber to operate in a neutron spectrometer to monitor for high-energy neutrons. Chiles was assigned to this research and development project. He^3 gas was selected as the detection media for the neutrons but it is a very expensive gas. Therefore the objective was to reduce the gas volume outside the sensitive part of the chamber to a minimum. Conventional proportional chambers use "Field tubes" to cover the center electrode at each end to identify a uniform sensitive volume in the center. This causes a lot of dead gas space at each end so an attempt was made to reduce that. The sensitive volume was to be equivalent to a 10,000-cc chamber. Therefore, the physical size of the chamber had to be small enough to locate near the experiment. So the objective was to design a chamber volume of 1000cc and pressurize it to 10 atmospheres of He^3 gas. This meant that the outside housing had to be welded stainless steel with ceramic insulators and no organic material at all inside the chamber. An experimental chamber was designed to use an alpha source inside and a less expensive gas such as argon to determine proper size for the center electrode. A special thermal pump had to be designed using liquid helium to transfer the He^3 gas to the completed chamber to 760-cm Hg absolute pressure. This allowed evacuation and gas filling without any chance of gas becoming contaminated from organic materials that would be present in a conventional vacuum pump system. The chamber performed fine using an amplifier with long pulse rise time due to the slow pulse collection time in the high gas pressure in the chamber. It maintained its integrity well and was being used when I retired.

Several scientists in Analytical Chemistry and Metallurgy Divisions were doing X-ray scattering and diffraction experiments that required exceptional pulse height resolution and Richard Zedler designed many X-ray sensitive gas proportional chambers for their experiments. Zedler did a lot of experimental research using different gas mixtures of inert gases such as argon, krypton and xenon to determine proper mixture to achieve best pulse height resolution for X-ray proportional chambers. He published many scientific articles about his gas research. Zedler also did a lot of early development work on Germanium and silicon surface barrier diode detectors for high-resolution alpha detection. With his knowledge and experience in chemistry, he made a significant contribution to the development of this new type detector especially on chemical etching to prepare the oxide surface barrier on the front side of the metal wafer.

The Operations Division requested the Detection Section to design an in-line large area alpha scintillation detector to monitor process waste lines and surface run off water in the small streams around ORNL area. This had never been done mostly because the range of alpha particles is so short in water. It is true that alpha particles do not travel very far in water so the depth of water from which these particles will enter the window of a detector is less than 0.001". The window material between the water and surface of the (ZnS) zinc sulfide scintillator reduced the depth even more. This dictated that we have a large sensitive area in order to have any appreciable detection geometry. From research on thickness of light pipes for large area alpha scintillation detectors usable with inexpensive 2" diameter photo-multiplier tube, a 1" thick and 6" diameter Methyl Methacrylate (Lucite) was used for the light pipe and substrate for the ZnS phosphor. Aluminum coated Mylar film 0.00025" thick was used for the window to protect the scintillator from the water. The housing and the water lines to and from the detector were made of stainless steel to allow easy decontamination in case that was needed. Ten such monitors were installed at ORNL. Calibration sensitivity measured one nano-Curie (1×10^{-9} Ci/ml) of U^{238} per milliliter of solution.

The United States Corps of Engineering requested assistance from I& C and Isotopes Divisions at ORNL to explore the idea of tracing sand movement along eroding beaches. This project was known as "Radioactive Isotope Sand Tracing", (RIST). The idea was to use sand granules spiked with a short half-life radioisotope (^{198}Au which has $T_{1/2}$ of 2.7 days and emits gamma photons of 412 kev) and dumped in the surf along certain beaches and monitor the movement of the radioactive sand with a radiation detector drawn behind a motorboat. Hugh Brashear was assigned to design the detector system and managed the operation of the equipment. Brashear also was the main person to analyze the data collected. Brashear is shown in the next figure, as he worked in the lab to ready the large computer system for the RIST tasks.

Hugh Brashear is shown in his lab, as he worked to ready the computer system (shown behind him) for the RIST tasks.

The detector system had to be rugged, water proof and protected from debris in the salt water. It consisted of four scintillation detectors mounted side by side in a wire mesh cage. The scintillators were 2" diameter x 2" long CsI(Na) optically coupled to 2" diameter photo-multiplier tubes and sealed water tight in a housing. The orientation of the detectors had to maintain a sensitive direction downward in order to detect the sand along the bottom of the ocean as it was being towed behind the boat. This was a challenging engineering project to accommodate these requirements. Roy Simpson in the Data Processing & Analysis Group assisted in computer analysis of the data. Information derived from these experiments assisted the Corps of Engineering in establishing barriers along beaches to reduce the amount of sand erosion occurring.

T. V. Blosser in the Neutron Physics Division was performing experiments for National Aeronautics and Space Administration, (NASA), to determine radiation exposure to astronauts during space flights from cosmic radiation and secondary radiation caused by cosmic radiation interacting with nuclei in the upper atmosphere. Blosser's experimental model consisted of a Methyl Methacrylate (Lucite) sphere filled with water and a tissue equivalent ionization chamber that could be remotely positioned anywhere inside the water filled sphere. Marion Chiles did the design and fabrication of the tissue equivalent ion chambers. The electrodes were fabricated from Methyl Methacrylate. Coating the surfaces of the plastic anode and cathode with a carbon filled conductive coating made the electrical contacts. The chambers were evacuated and backfilled with nitrogen gas.

The Radiation Detection Section was responsible for specifications to purchase commercial radiation detectors that were stocked in ORNL Stores as replacement parts in many radiation measuring instruments. The instrument repairmen reported that they were receiving GM tubes from ORNL Stores that were not operating properly. This prompted an investigation of the tubes in stores and approximately 50% of them were not meeting specifications. Therefore, a testing facility was prepared and all new GM tubes purchased for stock at ORNL were tested before they were accepted on purchase orders. Those not meeting specs were returned to the manufacturer for replacements. Several staff members were involved with this project: Hugh Brashear prepared the computer program, Clint Miller, Charles Fowler and James Ramsey assisted in fabricating the test facility and Marion Chiles analyzed the test data. Tube checker Bill Clowers tested the tubes and collected the data sheets. Chiles prepared reports for Purchasing Division about which tubes were accepted and those rejected. This improved the quality of performance of the many monitoring instruments throughout ORNL.

John Mihalczo in the I&C Reactor Controls Section was developing an electronic system to determine fissile assay of fissionable material. His system, (Source-Driven Sub-criticality Measurements) required a fast collection time fission chamber for his time of flight electronic signal. ^{252}Californium which has 3% spontaneous fission, was selected to provide a trigger signal from the fission chamber to start the time of flight circuit. Therefore this signal required a pulse with rise time which is congruent with the rate of charge collection time in the fission chamber of a few nano-seconds. In order to achieve these fast pulses in the detector they were filled with methane gas which has a very fast electron velocity and allows a short charge collection time. Frank Gillespie provided the first ^{252}Cf fission chambers. However, these chambers were sealed with epoxy and after a few months the methane gas became contaminated from organic materials used in assembling the chambers and the pulses were not fast enough thus required re-filling with high purity methane. Following this experience it was decided to design a fission chamber with no organic material used. All metal parts were made from stainless steel and weldable ceramic insulators were used.

At this time Chiles was brought on the scene to design a chamber that would maintain gas purity for much longer time. Also more strenuous requirements for containing ^{252}Cf were implemented at ORNL, thus demanding that ^{252}Cf be double contained in welded stainless steel housing. So, this brought on more complicated design and fabrication techniques. A new design was developed to meet these requirements, maintained gas purity for the usable life of the ^{252}Cf and performed well. As Mihalczo planned experiments in different fissile material configurations, fission chamber shapes and sizes were changed. The assembling and welding of these chambers had to be done inside a "hot cell" with shielded optical window and hand operated manipulators. This required special perception in designing the chambers to enhance this remote operation. The most difficult design to make was one that would fit inside LWR fuel element spacing of 0.375" diameter. Of course locating availability of small, weldable ceramic feed-through connectors was critical for the success of meeting this size limitation.

In the previously mentioned experiments that Mihalczo was performing to measure fissile material using his source driven technique, multi-energy neutron detectors were required to detect the secondary neutrons in the surrounding fissile material. A dual scintillator detector mounted on a single photo-multiplier tube was designed. This detector was capable of counting thermal neutrons, high energy neutrons as well as gamma photons, separately. Chiles designed the detector, Richard Todd provided a current-sensitive preamplifier required to separate the different rise time pulses from the different type radiations and of course Mihalczo used the electronic system in his experiments. A patent was issued from the U. S. Patent Office for this design.

In the mid 1960's, Instrumentation and Controls Division was greatly involved with the developing NASA program to send an astronaut to the moon. One of their prime investigations was to determine the natural elements in the surface of the moon and if any radioactive material was present. Vern A. McKay, Ted Sliski and Richard Wintenberg in I&C Division were assigned to design and manage the procurement, fabrication installation and checkout for operation of a "low-level Gamma-Ray Spectrometer System" for the Lunar Receiving Laboratory (LRC) located at Johnson Space Center, Houston, Texas. Also P. R. Bell was liaison between this group and NASA. McKay was responsible for the design of the Scintillation Detectors as well as for the radiation shielding required for low level background. Sliski was assigned to engineering and assembling the shielding cubicle. Wintenberg was assigned the task of designing and engineering the complicated electronic system for all the coincidence and anti coincidence circuits involved for the multi-detectors array. An engineering model was made and installed in Analytical Chemical Division at ORNL previously to eliminate problems that might occur in the final design to be installed in LRC. This was a long assignment but was very successful enabling NASA to analyze the moon rocks for their qualitative and radioactive contents.

A group of physicists and engineers from several National Laboratories were exploring the existence of high energy particles. Hans Cohn in the Physics Division at ORNL contacted Hugh Brashear about designing a large multi-wire gas proportional chamber to investigate physical characteristics of such high energy particles. Brashear designed the chamber and directed the development and fabrication of it. A special clean room was required for assembling and working on the chamber because any fiber or dust particles could not be present inside the chamber. One of the obstacles that required developing special techniques was to determine the proper tension in each wire as it was installed. This was accomplished by first calculating theoretically how much tension required and then developing a gauge to hold the wire at the proper tension while being soldered in place. I understand there were three parallel planes of wires stacked on top of each other. Each plane contained the equally spaced parallel wires. Each plane of wires was rotated 120 degrees from the adjacent plane. An important objective for using this chamber was to determine the location that a high energy particle made a hit (being detected in the chamber) and this arrangement helped in determining the location of each hit. This also required that the spacing between adjacent wires was precisely equal. V. C. Miller and Bill Clowers assisted in the fabrication, assembling and testing the chamber. The final testing and installation was at Fermi National Laboratory in Chicago, IL.

The Solid States Division of ORNL was doing experiments in Small Angle Neutron Scattering at High Flux Isotope Reactor (HFIR). They requested a large area thermal neutron sensitive multi-wire position sensitive proportional chamber to locate near their target being bombarded in a collimated beam of neutrons from the HFIR. W. T. Clay modified a previous model chamber to have 20" X 20" sensitive area which was much larger than previous chamber. The chamber was filled with ^{3}He gas, which has a high cross section for interacting with thermal neutrons. By using state of the art position sensitive electronics, the location that a neutron interaction occurred in the chamber could be determined. C. E. Fowler assisted in developing this chamber and designed and built a system to recover the expensive ^{3}He gas ($130 per atmospheric liter) in case the chamber required disassembly for repair. This recovery system maintained high purity of the gas, which was extremely important, since the chamber contained about $10,000 worth of gas. A major concern during assembling the chamber was cleanliness of internal parts and especially no dust or fiber particle could be present. So, a special clean room with air filters was required to assemble the chamber.

As time progressed, funding for I&C Division began to come from sources other than DOE, known as work for others. The Navy's RADIAC Development program was one program in which the Radiation Detection Section participated. The Ruggedized Alpha Scintillation Detector (ZEALS) was developed under this program. Stephanie McElhaney was the primary developer with assistance from M.M. Chiles, James Ramsey and Martin Bauer. This development was submitted as an entry into the R&D 100 Awards competition in 1994. Another detector developed under the Navy RADIAC funding was a "Multi-Energy Neutron Detector for Counting Thermal Neutrons, High-Energy Neutrons and Gamma Photons Separately". Chiles was the primary developer for this detector. A patent was applied for but the BAR date ran out and a patent was not granted.

Also Chiles designed as developed five assemblies for Navy Calibration Centers to use for testing and calibrating their Health Physics survey instruments. A gamma source, beta source and a 100 sq. cm alpha source were placed inside each of these assemblies. Each source was located on a flat plate mounted inside a lightweight, hard aluminum carrying case similar to a small suitcase.

This document describes my recollection of the personnel that served in the Radiation Detection Section (later named the Sensor Systems Development Group) during the time I worked in the group from February 18, 1957 till I retired July 31, 1994 with the exception of a few that were in the group for short periods of time. A few that I can remember in that category were Russell Jones, Lee Hunt, M. A, Meacham. Also there are a lot of projects not mentioned because time and space will not permit a complete list of all contributions the Radiation Detection Section made toward the success of ORNL experimental projects and operations. The projects listed are not necessarily in chronological order as to when they were done.

Subsequent years of the Process Control and Instrumentation Section saw several groups formed under the leadership of Charlie Mossman, as Section Head. A group under the leadership of Bernard (Bernie) Lieberman and Baden Duggins continued the work for Chemical Engineering and Pilot Plant instrumentation. The Development work (later referred to informally as the "Invention-on-Demand") group was continued under the leadership of Tom Gayle, Ray Adams and Lou Thacker. A new activity, the Digital Computer Applications Group was begun by Ray Adams and continued under CD Martin and Jim Jansen. This group later became known as the Real-Time Systems Group, under Jim Jansen.

Refer to Appendix 1 for the ORNL Instrumentation and Controls Division Organization Charts.

Chapter 7

Development of the ORNL Personal Radiation Monitor

Robert H. Dilworth

Perspective of 1958 as seen today

As this is written, a fountain pen sized radiation monitor with proportional dose-rate indication would be a simple task, using modern surface-mount miniature components and integrated circuits. But in 1958, when the project was initiated, circuit designers in I&C Division were in the process of re-educating themselves from vacuum tubes to solid-state devices and circuits. Silicon transistors were in developmental status, and none were in miniature cases. The available transistors were germanium units whose collector-base leakage currents were an appreciable portion of desired operating currents for low power battery powered circuits, and these leakage currents varied strongly with temperature. Those transistors in miniature cases were intended for hearing-aid use, and had widely varying parameters. Ordinary electronic components such as capacitors, resistors, and transformers were just becoming available in miniature sizes, responding to a market for small transistor radios and for hearing aids. Accommodating these limitations in available components into the design process for the PRM was a major factor - one that is difficult to appreciate in the current era.

Concept

The genesis of the PRM concept followed a radiation excursion in 1958 at the Y-12 plant, in which workers were alerted by area monitors. However, they had no way to choose an evacuation route that minimized exposure. Radiation detection devices available for use on the person were usually film badges or pocket quartz-fiber dosimeters which gave only accumulated dose information, and that only in retrospect. The desirability for personal instrumentation giving an immediate proportional indication of dose rate was made clear in this incident.

This event challenged C. J. Borkowski, director of I&C Division. He also headed the Instrumentation for Chemical Research Section, and at that time had his office in 4500 Building with that small group of I&C development engineers who served special needs of the Chemistry Division. This group also worked at state of the art improvements in radiation detection, measurement, and spectrometry. In discussions with this group, Borkowski expounded on the need for improved personal monitoring, suggesting that new solid state electronic circuits should be amenable to meeting the size and portability requirements.

These discussions brought up the recent availability of small geiger counter tubes having the needed sensitivity range, and operating with 500 volts dc rather than previous units requiring 900 volts or so. The challenge he outlined was to provide this voltage in efficient battery-powered miniature circuits, together with suitable visual or audible indication of dose rate.

At this time, R. H. Dilworth made Borkowski aware of work he had done in making small efficient sound generators using hearing aid components. In this work, done primarily in his home workshop as part of his re-education from vacuum tube to solid state circuits, a germanium transistor audio oscillator drove a hearing aid earphone, to which was added a resonant air column. The resonant column made a remarkable improvement in the level of sound, filling a room, though utilizing milliwatts of power and being physically small. Dilworth constructed a demonstration sound maker consisting of a mutivibrator having a period of about a second, in which one of the two germanium transistors cycling on and off was also an audio oscillator driving a hearing aid earphone coupled to a resonant air column. Since the air column, the earphone, and the small 9 volt battery were all in cylindrical shape, he chose a phenolic tube of 3/4 inch diameter and 6 inches length as a housing for the circuit. It was similar to a fountain pen. When energized by a small switch in the base of this tubular assembly, the device made continuous beeps at a noticeable sound level. See photograph - Fig. 1. It was found recently in Dilworth's attic. When the long defunct battery was replaced with an external power source, it came to life and beeped weakly.

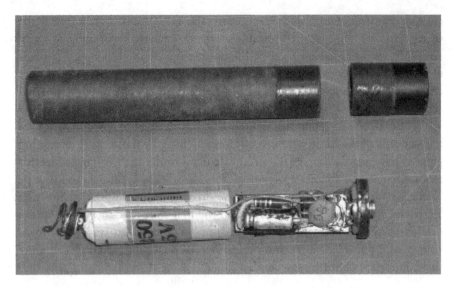

Fig. 1 The tubular experimental beeper that gave Borkowski his concept of a fountain pen sized PRM in 1958.

It was at this point that Borkowski conceived the fountain pen shape as the desired configuration for a proportionally indicating personal radiation monitor. Although that shape was somewhat inconvenient as a mechanical basis for solid state circuits, he insisted that it should be the final configuration, and he commissioned Dilworth to start work on the project.

Development

The full details of the evolution of the PRM circuit are not suitable for this historical summary. Publication ORNL-3048 in the TID-4500 series issued in August of 1961 contains full circuit information. However, the salient innovations that made the PRM possible can be summarized (see Fig. 2 for circuit references).

The predominant innovation was the use of the same blocking oscillator circuit for both the high voltage generation and as the source of the audible alarm sound. The instrument existed in low background level radiation for the majority of the time, and the very low geiger tube current in this condition allowed for a "keep-alive" very low repetition rate in the blocking oscillator high voltage generator, with consequent very low battery drain. However, when the instrument encountered higher radiation fields, a simple non-linear feedback scheme raised the repetition rate of the voltage generator to provide the required output current for the geiger tube. Fortunately, the practical repetition rates were in the audio frequency range. By choosing circuit parameters carefully, the repetition rate of this oscillator could be made to coincide with the desired audible alarm characteristics of a rising chirp rate with rising radiation level. This double use of the blocking oscillator made possible a very significant reduction in component count as a necessary criterion for keeping the circuit within the confines of a fountain pen sized case.

The dual use of the blocking oscillator was the first of three instances of the concept of multipurpose use of components. In the second, obtaining good voltage regulation versus current demand in the high voltage generator required that the flyback portion of the blocking oscillator waveform be "snubbed" to the same amplitude as the driven portion. Rather than using a separate zener diode for this function, the first shunt rectifier diode in the Cockcroft-Walton voltage multiplier was chosen to have an inverse breakdown voltage of the needed value, and performed the snubbing function, reducing the component count by one diode.

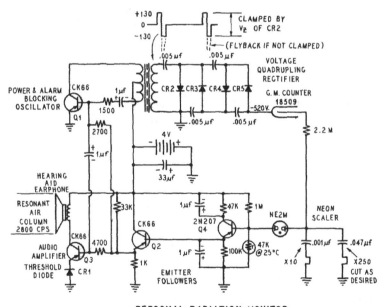

PERSONAL RADIATION MONITOR

Fig. 2 Circuit of PRM

In the third instance, some form of rudimentary scaler was needed to reduce the natural count rate of the geiger counter before driving the feedback circuit that made the oscillator rate proportional to radiation level. A simple scaler was devised in which the geiger tube pulses charged a capacitor until the ignition point of a neon lamp was reached, producing the desired reduced pulse rate. The neon lamp thus flashed at a rate proportional to radiation level, and provided a visual indicator of dose rate while simultaneously acting as a scaler.

A significant innovation in the PRM was unconventional use of the geiger tube, utilizing the averaged current output of the tube rather than its pulse rate as the measure of radiation level. Typical geiger tube use in radiation survey instruments used the tube pulse rate as the measure of radiation level. However, when these circuits encountered very high radiation levels well over the normal range of use, they would block, producing no reading, as the ionization events in the tube became essentially continuous and pulse detection circuits failed. A block in indication at dangerous radiation levels was intolerable for the PRM.

Fortunately, the small geiger tube employed, if provided with a steady high voltage source, would continue to produce an output current monotonically increasing with radiation level, finally flattening at very dangerous levels of several hundred r/hr. By sensing the geiger tube current rather than its pulse rate, blocking at high dose rates was eliminated. The non-blocking characteristic of the PRM was tested dramatically by a visit to one of the hot cells in the Isotopes Division, where sources were gathered around the test PRM until a gamma dose rate of 3 million r/hr was achieved. The PRM alarm sound never faulted.

The expected operating environment of the PRM would include temperatures near freezing around outdoor facilities in winter, and very warm temperatures when on the dashboard or in the glove compartment of a car or truck in summer. Obtaining this kind of performance with germanium transistors was a challenge since room-temperature leakage currents were an appreciable portion of operating currents. While still in the breadboard stage, the circuit was tested and modified until it would operate properly over a temperature range of 0 to 50 degrees C. A miniature thermistor in the input of the circuits determining oscillator repetition rate provided the needed temperature compensation.

The physical package of the PRM went through several evolutions before finalizing. A captive gold-plated battery contact scheme eliminated the common flashlight-type problem of poor contact through a screw-on end cap. The circuit components were epoxy encapsulated for protection. The stainless steel case and its method of fabrication were the work of I&C staff member R. J. Fox. The construction of some of the breadboards and of the many prototype instruments was done by Engineering Aide J. L. Hutson.

Prototype Evaluation

Once the circuit parameters were established, five prototype PRM instruments were built in simple enclosures resembling fountain pens. These were used primarily as demonstrators, since reliability was not yet established for field use. Since the small geiger tube had thin wall construction, it could be tested by use of a beta radiation source. A suitable portable safe source was developed in conjunction with Isotopes Division. A stainless steel capsule having a thin end window contained evaporated Strontium-90 material, and was mounted inside a cigar shaped aluminum holder. A lead-lined end cover could be unscrewed to reveal the thin window, and beta radiation equivalent to about 2 r/hr gamma radiation would activate the PRM. These demonstration units were carried about for many weeks and used to detect various radiation sources in use around ORNL as well as for demonstrations with the portable beta source. At the conclusion of these first evaluations, the results were such as to justify building a larger quantity of the instruments for further evaluation. Also at this time, a circuit modification was made allowing for selection of a normal or a low sensitivity radiation level range.

Thirty additional prototype instruments were then constructed, but not before devising a program to identify and solve problems a manufacturer might encounter. Because of the widely varying parameters of the germanium transistors intended for hearing aid use, quantities of the units were tested and divided into categories of *"beta"* the small-signal common emitter gain, as measured by a widely used transistor tester. Based on these tests, units were assigned to specific circuit locations unless rejected. Similarly, a few resistor values needed to be hand selected in a test breadboard to compensate for transistor variations. These kits of matched parts were then used in prototype construction.

By the time the 30 sets of parts had been assembled, the fountain pen sized case design was finalized and 30 cases fabricated. Thus the 30 prototypes were of a design sufficiently firm that successful experience with them could allow purchase of a quantity of instruments for ORNL and the release of full manufacturing information to prospective instrument companies. The 30 prototypes were placed in use within ORNL, and after some months the results were satisfactory and the design was considered successful.

Manufacturability Factors

Since both ORNL needs and a possible public market made manufacture of quantities of the PRM likely, attention was given to those factors that might better insure successful manufacture. The experience with selection of components to form kits of matched parts was analyzed, and a parts kit selection breadboard was designed incorporating switch selection of the few resistor values that had to match the specific transistors of an individual parts kit. Drawings were prepared to allow duplication of the selection breadboard. Then the PRM design was documented into a full set of specifications and drawings entitled Q-2041, for use in potential future ORNL procurement. In addition, a similar complete "CAPE" package of information was prepared for the technical information distribution service of the AEC.

Fig. 3 shows the finished instrument

Procurement

To facilitate procurement of a quantity of the PRM instruments for ORNL use, three instrument manufacturers who had expressed interest were chosen for further involvement. Full sets of all parts needed to build several units along with the full manufacturing instructions were provided these three. On evaluation of their samples, only one was fully acceptable, and became the source for a subsequent production order.

Publicity

The PRM differed from most I&C development projects in that it had a function and appearance that was immediately interesting to the general public, which was just coming into an awareness of miniature transistorized products. This fact was recognized by the public relations office of Union Carbide Nuclear Corporation, the operating contractor for ORNL. With the assistance of I&C staff, that office prepared publicity releases which it disseminated widely. Newspapers all over the country printed short articles. Time magazine devoted a full column to the PRM, in its issue of April 14, 1961, including the picture as shown above in Fig 3 (in the hand of this writer), and a probably fictitious quote from an unidentified "worker" saying, "it tells us when to run like h---".

Many manufacturers of PRM components, notably the transistor and battery makers, featured the device in advertisements. This publicity effort by UCND was very successful in communicating a positive image of ORNL to the public.

An Appreciation

This writer would like to take advantage of this forum to state an appreciation of the late C. J. Borkowski, not only for his concept and support of the PRM project, but also for his positive leadership of I&C Division. But his greatest contribution to ORNL was his vision of instrumentation as a discipline worthy of inclusion in the research mission of a national laboratory. He worked to provide funding other than specific work orders for innovative instrumentation that expanded the state of the art. Such funding was available when the PRM project was launched, and it also provided for the subsequent creation of the Basic Instrumentation Section in I&C Division. Subsequent administrations have not supported his vision, even to the present point of disbanding the division. ORNL is the poorer.

ABSTRACT

Developed in 1958 by C. J. Borkowski and R. H. Dilworth, the ORNL Personal Radiation Monitor was a fountain-pen sized radiation detection instrument providing a visual and aural output proportional to radiation dose rate over a range of from background radiation to dangerous levels. In background radiation, the PRM would chirp once every minute or so, assuring the user that it was operating. The chirp rate would increase in higher radiation levels until a continuous loud scream was reached at a few r/hr. Worn in a shirt pocket and operating continuously for a month from a self-contained battery, the PRM could warn workers of unexpected radiation exposure and help them choose a route of safe evacuation. The design accommodated quantity production. A number of the units were purchased for ORNL use, and they were available for some time from commercial suppliers. The PRM development is of historical interest as an early example of miniaturization of instrumentation at a time when silicon transistors were experimental devices and when most of the small components came from the hearing aid industry.

PRM Anecdotes from Bob Dilworth

When I graduated from college with a degree in Engineering Physics in 1952, I had never seen a transistor. The first I encountered was in an Air Force laboratory where I was stationed. It was a Raytheon point contact germanium device. I immediately took possession of it, and based on what meager technical information I had at hand, built an audio frequency oscillator on the headband of an old set of earphones. When turned on, it produced a steady tone at a modest sound level. My comrades and I had great fun hiding it in desk drawers. When the occupant of that desk would ask if we heard anything humming, we would, of course, say no.

Later when I was teaching myself solid state circuits about 1957 in my home workshop, the first thing I tried to build with a germanium junction transistor was another audio oscillator, this time driving a hearing aid earphone. The sound level was low, and I stumbled onto the use of a resonant air column attached to the earphone to greatly enhance the sound level. This work led to the tubular beeper mentioned in the Concept section of this history, from which Borkowski took his concept of a fountain pen sized personal radiation monitor.

When writing about that tubular beeper, I went upstairs to my attic to see if that old device was still in a box labeled 'Keepsakes'. Sure enough, it was still there (now 46 years old. The battery had long since expired.) But getting clip leads out and connecting an external battery to the old relic, I was delighted to hear it first tick a bit, and then as the old electrolytic capacitors re-formed themselves, it began to emit somewhat feeble beeps. Aging of the old components had shifted the beep frequency somewhat away from the tuning of the resonant air column, but none-the-less, it still beeped!

One of the more interesting phenomena told me after some PRMs were in use was that when traveling by high-altitude jet aircraft, the increase in background radiation level from cosmic radiation at such altitudes was such that the chirp rate of the PRM became noticeable to others. For that and other occasions when it was desirable to turn the PRM off, users began to carry small discs of plastic sheet to slip between the battery end and the contact to turn the PRM off.

I left ORNL shortly after the PRM project was complete, to go to private industry. Fifteen years later, I returned to ORNL in management

positions. But I found I was yet remembered as 'the guy who invented the pocket screamer'. I felt at home again

Editor's note: The PRM, in its original form, (shown disassembled on a display board, Fig. 4) was used for several years by ORNL staff. However, it required careful selection

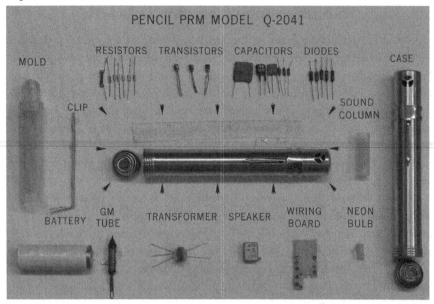

Fig. 4 Display Board showing the physical components of the PRM

of components for manufacture, and, as time passed, its repair required time-consuming application of dental drills and fine mechanical work to replace failed components within the epoxy-sealed cylinder. By this time also, better (silicon) transistors became available, and the circuit and its form was modified to a more easily manufactured one resembling a cigarette pack. This later model (shown in Fig.5) was also more easily repaired, and was used at ORNL for many more years.

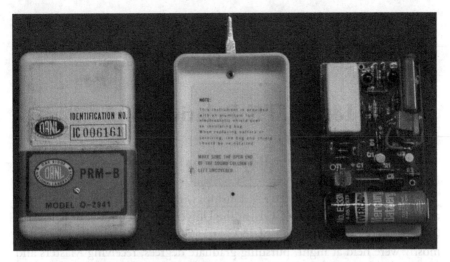

Fig. 5 Cigarette Pack version of the PRM

Publications List

ORNL Personal Radiation Monitor

Instrumentation and Controls Division Annual Progress Report for Period Ending July 1, 1959 ORNL-2787 Page 55

Instrumentation and Controls Division Annual Progress Report for Period Ending July 1, 1960 ORNL-3001 Page 22 This article contains the fullest description of the circuit operation.

Transactions of the American Nuclear Society, 1960 Winter Meeting, Volume 2 Number 2, December 1960 Page 454

State of the Laboratory ñ 1960 Alvin M. Weinberg ORNL Central Files Number 60-12-112 Page 8

Personal Radiation Monitor R. H. Dilworth and C. J. Borkowski ORNL 3058 in TID-4500 Series Issued August 28, 1961

Oak Ridge National Laboratory Review Volume 11, Number 3, Summer 1978 Issue Featuring ORNL and Industry Page 12

U. S. Patent No. 3,015,031, Personal Radiation Monitor, December 26, 1961 R. H. Dilworth and C. J. Borkowski

Chapter 8

I&C/UTK Relationships

Ray Adams

The I&C Division fostered relationships with several of the departments of the College of Engineering (COE) at the University of Tennessee at Knoxville (UTK). Members of the staff of I&C attended graduate school classes that mostly were held at night, pursuing graduate degrees, receiving Masters and PhD degrees at UTK. In addition, several of the Professors in various COE Departments, worked part time at ORNL, in the I&C Division.

Relationships With Electrical Engineering

UTK COE staff members from the Electrical Engineering Department either worked in the I&C Division under a contract that was set up with the Electrical Engineering Department under its chairman Dr. Joseph Googe (for many years), or held their own consulting contract. They were Joseph M. Googe, Igor Alexeff, Vaughan Blalock, Robert Bodenheimer, Donald Bouldin, Ralph Gonzales, Ed (Eldridge) Kennedy, Frank Pierce, Carl Remenyik (Carl actually was in the UTK ES&M Department, but worked under Googe's contract), James Rochelle, Robert Rochelle, Fred Symonds, & Charles H. Weaver. Some of the UTK faculty participants taught mini-courses for the staff of I&C at ORNL. Among the courses were Transistors and Application Specific Integrated Circuits – in addition to standard Graduate courses.

Relationships With UTK Nuclear Engineering

UTK COE staff members from the Nuclear Engineering Department often worked on specific projects in the I&C Division. Their work was generally covered by the research program on which they worked at any one time. Over the years, faculty and students in the Nuclear Engineering Department included Pete Pasqua (First head of NE at UTK), Tom Kerlin, Ralph Perez, Belle Upadhyaya, & Jim Robinson,

The Measurement and Controls Engineering Center

Dick Anderson and Ray Adams

Charlie Mossman was the inspiration for what eventually became the Measurement and Controls Engineering Center at UTK. Charlie for years had complained that there were no university programs that had I&C degrees. The result was that new hires into the I&C Division spent about three years coming up to speed to qualify as I&C engineers. Charlie wanted to push for an I&C degree program at UT.

In our initial conversations with Dean Snyder and others at UT, it became obvious that establishing an I&C Department at UT that would grant degrees in I&C Engineering would meet with almost insurmountable obstacles. First of all, establishing a new Department in the College of Engineering would create a new overhead. As a result it was highly improbable that the powers that run the University would be likely to support the idea of a new Department. Second, I&C Engineering was (and is) not a widely recognized academic discipline. On the plus side, faculty members in the Engineering College were favorably disposed to developing some kind of inter-departmental program in I&C studies.

About this time we heard of a relatively new program of the National Science Foundation for University-Industry Cooperative research programs. Several of us (Myself, Ray Adams, and one or two UT profs) went to Washington to talk with Alex Schwartzkof the program manager, to explore the ideas of a cooperative research center in I&C. Alex laid out the ground rules concerning the kind of market research we would have to carry out to show broad enough interest for NSF support.

On returning to TN, Bill Snyder enlisted the aid of Henry Guigou – of the UTK System Development Office, who was involved in generating contributions to the University from industry. One of Henry's strengths was acquaintance with a number of UT alumni in managerial positions that could be called on to sponsor the idea in their companies. This resulted in a number of trips to places like TN Eastman, the Alcoa Research Center, Brown & Root, Shell Development Labs, the General Electric Research Labs, and others. We also visited one or two of the established NSF research centers to learn how they put together their program.

These efforts were eventually successful and the Measurement and Control Engineering Center (MCEC)was established at UTK, with support from ORNL. The MCEC was established in 1985 and its first director was

Emil (Bud) Muly. By the year 1998 it had established an affiliation with Oklahoma State University, and had partnershipped with more than 20 industrial supporters, including, Alcoa, 3M, Allied Signal, Dow Chemical, Union Carbide, Eastman Chemical, Exxon, Universal Oil Products, Hercules Powder, Glaxo Welcome, (and others). These partners had royalty-free access to inventions and other results of the graduate student/faculty research work at the center.

The Center was set up with an Industrial Advisory Board from industry, who judged proposals for work, as submitted by industry. This Industrial Advisory Board advised the MCEC on all aspects of the Center's work, from strategic planning to research selection. Some of the proposals were for proprietary work that could not be shared with all of the Board Members. This arrangement was somewhat foreign to the normally open aspects of ORNL work, but was nevertheless handled within the arrangements established for the MCEC. One of the crucial aspects of the MCEC was significant funding by the National Science Foundation (NSF) starting around 1986. The NSF required annual reports of the work of the UTK MCEC.

The first director of the MCEC was Emil (Bud) Muly. Later directors were Arlene Garrison, John Coates, Eric Sundstrom, Rich Jendrucko, and Kelsey Cook. The MCEC closed in 2006. Some of the COE faculty members involved with the MCEC were: Mike Roberts (ECE), Charlie Moore (ChE), Bill Hamel (MechE), Marion Hansen (MSE), Belle Upadyaya (NucE), and Jack Wasserman (Eng. Sci & Mech).

Chapter 9

Small Diameter Thermocouple Research

R. L. Anderson

Physical Considerations

Large-scale reactor safety experiments in the Engineering Technology Division small diameter (0.020-inch), metal-sheathed thermocouples. These thermocouples were incorporated into scale model fuel-rod simulators, which were high energy density electrical heaters used to simulate fuel rods in nuclear reactor cores. A single experiment could employ several hundred of these thermocouples, which required thousands of feet of thermocouple materials. One of these experiments, the Core-Flow Test Loop (CFTL), a high temperature helium loop for tests of gas-cooled fast reactor components and verification of thermal-hydraulic data, required much higher temperature service than previous experiments, and much of the research described below was conducted to characterize the small diameter thermocouple materials up to ~1370 °C.

The small dimensions of the thermocouple wires, sheath and insulating spaces between them resulted in a greatly increased sensitivity of these sensors to physical and chemical effects on the several components. These small diameter thermocouples (T/Cs) behave differently from larger diameter thermocouples in their mechanical , electrical, and chemical properties. The average wire size is on the order of 0.003 inch, or about the diameter of a human hair. The typical spacing between the wires is on the order of 0.002 inch and is filled with MgO insulation.

The electrical properties of the MgO insulation can be seriously degraded by the adsorption of water vapor due to the hygroscopic nature of MgO if sufficient precautions are not taken to prevent the exposure of the MgO to moist air. Figure 1 shows the behavior of the insulation resistance observed in small diameter T/Cs.

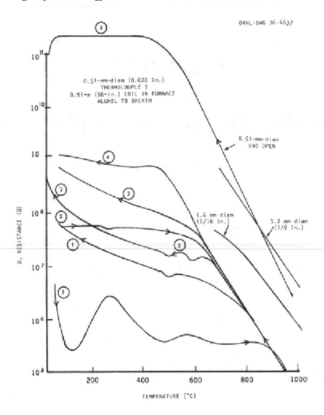

ORNL-DWG 76-4637

Fig. 1 Insulation resistance of a 0.51-mm-diam. metal sheathed thermocouple sample during repeated thermal cycles to 1000 °C. The numbers indicate the sequence of thermal cycles from 1 to 5 and the arrows indicate either the heating or cooling part of the cycle. The insulation resistance increased with each thermal cycle.

The initial heating shows a large dip around 100 °C, indicating the release of sorbed H2O. Subsequent cooling from ~1000 °C resulted in an increased insulation resistance through a combination of pumping (forcing the water vapor to the cooler sections of the thermocouple material) and through chemical reaction of the H2O with the thermo elements or the sheath. Repeated thermo cycling tended to increase the insulation resistance, which in this case, finally reached a limiting value of ~10^9 ohms. Another sample, with the end of the sheath left open to allow the water vapor to escape, after being heated to 1000 °C and cooled back to room temperature attained a much higher insulation resistance (<10^{11} ohms) until the temperature of the sample fell below 100 °C. As the temperature fell below 100 °C, the open end of the thermocouple sample allowed water vapor to reenter the thermocouple and the insulation resistance decreased.

At high temperatures, the exponential increase of the conduction of the insulation with increasing temperature can result in temperature measurement errors in three ways: 1. the increased conduction between the thermo elements in the hot zone in the temperature profile shown in Fig. 2.1, results in shunting and a decrease in the measured thermocouple extends *through* a temperature profile as shown in Fig. 2.2, and when the peak temperature of the profile is above ~1000 °C, the thermocouple output tends to reflect this peak temperature rather than the temperature of the measuring junction; 3. D.C. leakage currents on the sheath of the thermocouple (Fig. 2.3) can leak onto the thermocouple circuit and produce large positive or negative errors depending on the direction of the leakage current. This is because the resistivities of the two thermo elements are, in general, different. (In Chromel versus Alumel, for instance, the ratio of the resistivities is about 2:1.)

ORNL–DWG 77–14504

AT HIGH TEMPERATURES ELECTRICAL
CONDUCTION BETWEEN THE WIRES AND
BETWEEN THE WIRES AND SHEATH CAN CAUSE:

1. SHUNTING OF THE THERMOCOUPLE
   SIGNAL (– ERRORS)

2. CREATION OF VIRTUAL JUNCTIONS
   (+ ERRORS)

3. ALLOW ELECTRICAL CURRENTS ON THE
   SHEATH TO LEAK INTO THE THERMO–
   COUPLE CIRCUITS (± ERRORS)

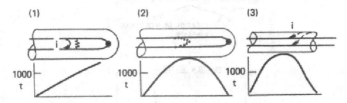

Fig. 2.1, 2.2, & 2.3, The electrical shunting between the thermocouple wires and the sheath depends on the profile of the temperature gradient.

Mike Roberts and Tom Kollie developed an analytical model based on three-wire transmission line theory which accurately predicts the errors in any or all of the mechanisms described above[11]. The model sums increments segments in which the resistances of the wires and sheath, the conduction between the wires and between the wire and the sheath can be represented by resistances, and the emfs generated in the various segments are represented by voltage sources, all as a function of an applied temperature profile (such as that shown in Fig. 2.2).

ORNL–DWG 77–14505

**ERRORS CAUSED BY ELECTRICAL SHUNTING
AND ELECTRICAL LEAKAGE CAN BE
CORRECTED AND/OR INTERPRETED BY USING
AN ANALYTICAL MODEL DEVELOPED FOR
CFTL SMALL DIAMETER THERMOCOUPLES.**

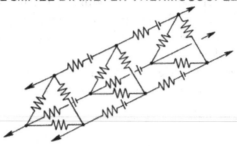

*Fig. 3, Diagram of the analytical model used by Kollie and Roberts to calculate
electrical shunting effects in metal sheathed thermocouples.*

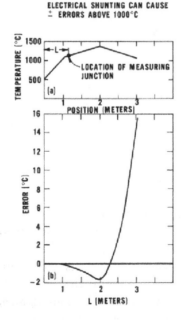

*Fig. 4, Results of a sample calculation using the Kollie-Roberts model. The
temperature profile used in the calculation is indicated in the upper graph. The
resulting temperature error is given in the lower part of the figure.*

116

The results of a typical calculation is shown in Fig. 4 for the temperature profile given above, and the temperature error shown is that which would result as the thermocouple was inserted through the peak of the temperature profile. This behavior was confirmed by experiment, and the model correctly predicted the measured errors within 10% in all cases.

ORNL–DWG 77–14506

CHEMICAL REACTIONS CAN AFFECT BOTH
ELEMENTS OF THE THERMOCOUPLE

REACTIONS MAY TAKE PLACE BETWEEN THE
THERMOCOUPLE WIRES AND:
1. THE INSULATION
2. IMPURITIES IN THE INSULATION
3. THE ATMOSPHERE INSIDE THE SHEATH
4. THE SHEATH

RATES OF CHEMICAL REACTIONS, AS A
GENERAL RULE, DOUBLE FOR EACH 10°C
RISE IN TEMPERATURE.

Fig. 5, Complex chemical reactions can occur among the various elements in a metal sheathed thermocouple.

Chemical Considerations

Exposure of small diameter thermocouples to high temperatures can cause extensive changes because of chemical reactions within the sheath. Fig. 5 illustrates the complexity of the system. Processes occur which result in a change in composition, and therefore change in the Seebeck coefficient of the thermoelements. Contaminates can diffuse from the insulation or through the insulation from the sheath into the thermoelements and components of the thermoelements may be preferentially removed through chemical reactions or vaporization.

This was demonstrated by some results of decalibration experiments run in the I&C Metrology Research and Development Laboratory (MRDL) on Type K (Chromel versus Alumel) thermocouples sheathed in Type 304 stainless steel as compared with Type K thermocouples sheathed in Inconel 600 (Figs 6 and 7).

Insert Photo: ORNL-DWG77-14508.tif

ORNL—DWG 77—14509

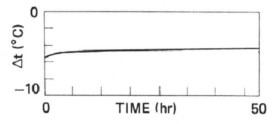

CHANGES OBSERVED IN TYPE K IN INCONEL—600 ARE MUCH SMALLER

Fig. 6, Results of experiments on exposure of Chromel vs Alumel thermocouples sheathed in type 304 stainless steel

Fig. 7, Results of experiments on exposure of chromel vs Alumel thermocouples sheathed in Inconel 600.

Other experiments were conducted on noble-metal thermocouples. The drift of Type S thermocouples, sheathed in Type 304 stainless steel, Inconel 600 and Pt-Rh alloys is summarized in Fig. 8, which shows the drift at 1300 °C over a period of 20 minutes. Type B thermocouples show a similar drift.

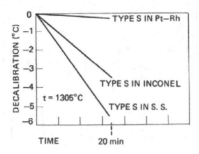

ORNL–DWG 77–14507

THE DRIFT OF THERMOCOUPLES IS STRONGLY
DEPENDENT ON THE SHEATH MATERIAL:

*Fig 8, Results of experiments on exposure of Platinum vs Platinum/10%
Rhodium thermocouples sheathed in various metals.*

Ion Microprobe Analysis

Cooperative efforts with the Ion Microprobe analysis group in the Analytical
Chemistry Division at ORNL provided some fundamental insights into the
decalibration processes. For instance, the noble-metal thermocouples were subjected
to a high temperature exposure in a tube furnace with the profile shown in Fig. 9.
At the bottom of the figure, the letters indicate locations along the length of the
thermocouple where samples were taken for analysis in the ion microprobe.

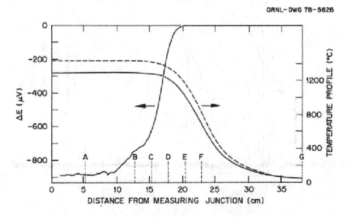

ORNL–DWG 78–5626

*Fig. 9, Decalibration experiments of a Type S thermocouple sheathed in Type 304
stainless steel. The furnace temperature profile is indicated on the right-hand axis and
the temperature measurement error is given on the left-hand axis. The letter along the
X-axis indicated positions where samples were take for microprobe analysis.*

119

Following, Figs 10 through 14 show the concentrations of Mg, Mn, Cr, Ni, and Fe in the composites of the analyses of all sections, A through G, of sample taken from a Type S thermocouple sheathed in stainless steel.

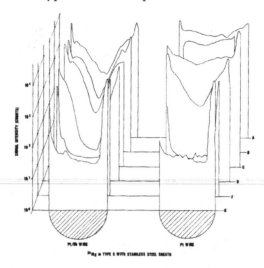

Fig. 10, Concentrations of Mg in a Type S thermocouple after high temperature exposure. The letters on the right-hand side of the figure indicated the corresponding position in the temperature profile shown in Fig. 9.

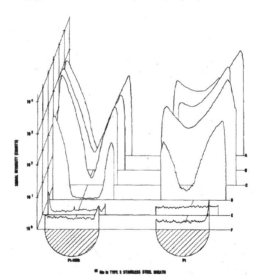

Fig 11, Concentrations of Mn in a Type S thermocouple after high temperature exposure. The letters on the right-hand side of the figure indicated the corresponding position in the temperature profile shown in Fig. 9.

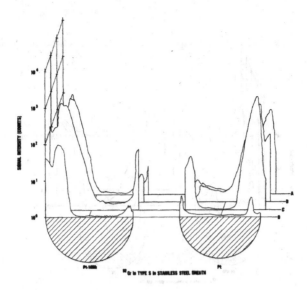

Fig. 12, Concentrations of Cr in a Type S thermocouple after high temperature exposure. The letters on the right-hand side of the figure indicated the corresponding position in the temperature profile shown in Fig. 9.

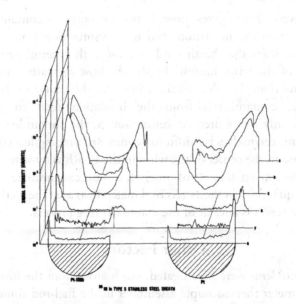

Fig. 13, Concentrations of Ni in a Type S thermocouple after high temperature exposure. The letters on the right-hand side of the figure indicated the corresponding position in the temperature profile shown in Fig. 9.

121

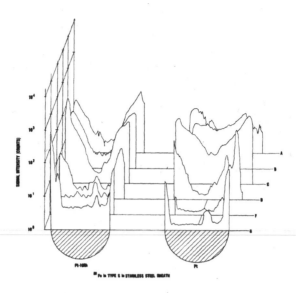

Fig. 14, Concentrations of Fe in a Type S thermocouple after high temperature exposure. The letters on the right-hand side of the figure indicated the corresponding position in the temperature profile shown in Fig. 9.

Qualitatively, these figures show: 1. two categories of contamination: a. contaminates from the insulation, and b. contamination from the sheath. (Contaminates from the sheath tend to *shadow* the thermoelements; that is, the edges of the wires nearest the sheath show a greater concentration of contaminate than the edges of the wires near the center of the sheathed assembly.); 2. contaminates from the insulation tend to produce a radially, uniformly distributed concentration; 3. the contamination of the thermoelements themselves is a diffusion controlled process (less contaminate tends to diffuse to the center of the wires in the Pt-Rh alloy wire than in the pure Pt wire in a given amount of time); 4. certain contaminates are present in significant quantities in the as-received material, as seen the section G. This latter contamination is a result of the manufacturing process.

Other Factors

Several problems were investigated which arose from the installation of the small diameter thermocouple assemblies in the fuel-rod simulators, the electrically heated fuel-rod simulators used in the large-scale experiments in the Engineering Technology Division.

The Blow-Down Heat Transfer (BDHT) program was brought to a screeching halt, when, on initial start-up, the thermocouples were found to be indicating temperatures in error by as much as 150% at about 150 °C. (Fig. 15).

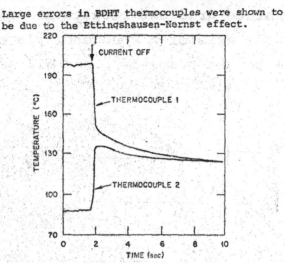

Fig. 15, The large temperature measurement errors that were observed in Type K thermocouples in BDHT fuel rod simulators disappeared when the simulator current was turned off.

Several weeks of intensive investigation in the I&C Division followed, which showed that the effect was due to the Ettingshausen-Nernst effect in Alumel. The Ettingshausen-Nernst effect produces an emf in a ferromagnetic conductor whose axis is normal to the plane of a magnetic field, which, in turn, is perpendicular to a temperature gradient (Fig. 16).

Application of a magnetic field $\perp$ to a temperature gradient results in an emf along the axis of a conductor placed normal to the plane of $\vec{B}$ and $\vec{\nabla T}$.

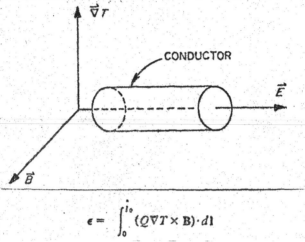

$$\epsilon = \int_0^{l_0} (Q \nabla T \times \mathbf{B}) \cdot d\mathbf{l}$$

Fig. 16, The Ettingshausen-Nernst effect is the emf produced when a conductor is located in a magnetic field perpendicular to the axis of the conductor while at the same time subjected to a temperature gradient also perpendicular to the conductor axis, but also at right angles to the magnetic field.

The laboratory investigation showed that the effect disappeared as the temperature was raised above the Curie Point of Alumel (about 152 °C). Subsequent trial runs in the BDHT loop confirmed this result. This work resulted in two papers: a general description of the results published in the Review of Scientific Instruments[12] and a subsequent measurement of the B versus H curves for Alumel as a function of temperature at the National Magnet Laboratory which was published in the Journal of Applied Physics[13]

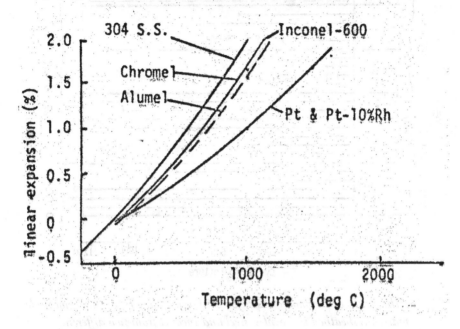

The thermal expansion of Chromel and Alumel matches Inconel, but not Stainless Steel

Fig. 17, Differential thermal expansion between the thermoelements and the metal sheath resulted in premature failure of the thermocouples after repeated thermal cycling except for Type K in Inconel-600. In this case the thermal expansion of the sheath and thermoelements was very nearly the same.

Initial experiments with prototype fuel-rod simulators for the Core Flow Test Loop (CFTL) resulted in nearly 100% thermocouple failures, particularly after thermal cycling. This was shown to be due to a combination of :

1. the differential thermal expansion between the Type 316 stainless steel sheath simulator cladding and the thermoelements of the Type K thermocouples (Fig 17), and

2. that recrystallization and grain growth in the thermoelements, particularly the Alumel, resulted in grain sizes comparable to the wire diameters. The grain growth tends to weaken the wires, so that the differential thermal expansion can pull them apart. A laboratory simulation of the effect showed that such failure is strongly dependent on the manufacturing process. Samples from one manufacturer failed, on the average, after only 30% of the number of cycles as those from a second manufacturer (Fig. 18).

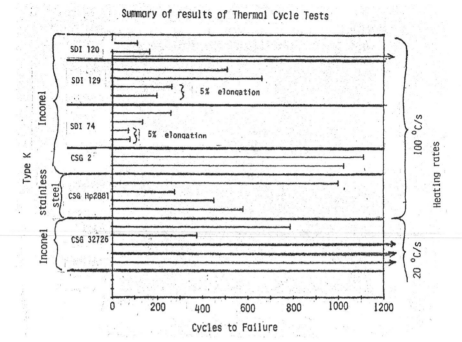

Fig 18, The results of thermocycle tests for 0.51-mm-diam Type K thermocouple materials sheathed in stainless steel and Inconel from two different manufacturers and different manufactured lots.

Almost every facet of the effects which can affect the performance of small diameter thermocouples was investigated over a period of about three years. Other areas that were investigated include: the tensile properties of small diameter thermocouples, i.e. the breakage of the thermoelements or sheath under stress; time response; new materials such as Nicrosil versus Nisil; and the use of boron nitride as an insulating material. Work continued on these topics and in particular a more extensive comparative study of Nicrosil versus Nisil and Type K thermocouple materials in vacuum and inert gas in both sheathed and bare-wire configurations.

Our experience with small diameter thermocouple materials provided many insights into the behavior of metal sheathed thermocouples in general, and resulted in improvement in the fundamental understanding of the processes which result in temperature measurement errors in thermocouple thermometry. Such results as these are typical of many of the studies in the I&C Division of ORNL. They illustrate the many (detailed) facets of the phrase "beyond the edge of technology". More complete descriptions of these efforts can be found in the following references:

REFERENCES

Anderson, R. L., and Kollie, T. G., "Accuracy of Small Diameter Sheathed Thermocouples for the Core Flow Test Loop," *ORNL-5401,* Oak Ridge National Laboratory, Oak Ridge, TN (1979).

Anderson, R. L., Lyons, J. D., Kollie, T. G., Christie, W. H., and Eby, R. E., "Decalibration of Sheathed Thermocouples," in *Temperature, Its Measurement and Control in Science and Industry,* Vol. 5, Instrument Society of America, Pittsburgh, PA (1982).

Mossman, C. A., Horton, J. L. and Anderson, R. L., "Testing of Thermocouples for Inhomogeneities," in Temperature, Its Measurement and Control in Science and Industry, Vol. 5, Instrument Society of America, Pittsburgh, PA (1982).

Anderson, R. L., and Ludwig, R. L., "Failure of Sheathed Thermocouples due to Thermal Cycling," in Temperature, Its Measurement and Control in Science and Industry, Vol. 5, Instrument Society of America, Pittsburgh, PA (1982).

Chapter 10

Emergence of the I&C Maintenance Management Department

Editor's Note: When the Instrument Department was formed within the E&M Division in 1949 (The Instrument Department [in 1953] became a major part of the Instrumentation and Controls Division), there was a separate Maintenance organization. It was then headed by Bill Ladniak and included all of the technicians and instrument mechanics (and a few machinists) that were shortly to become a part of the I&C Division. This group formed the bulk of the instrument maintenance mechanics at ORNL in those "early days".

D. R. Miller

Instrumentation and Controls work at ORNL began with a relatively narrow focus, that of creating new I&C technologies suitable for controlling and measuring the performance of one-of-a-kind nuclear reactors and processes in the associated chemical laboratories. Heretofore similar devices and technology had existed only in university research laboratories as handmade, somewhat fragile equipment. The equipment developed for ORNL would differ significantly from that required by university laboratories. Each measurement and control device would have to be able to be rapidly reproduced at other government laboratories and to sustain an unknown life expectancy. At the same time, the equipment had to be significantly more rigorous in detection sensitivity and precision of control than commercially available devices.

It became obvious, from the beginning, that a strongly integrated hierarchy of scientists, engineers, specialists and hourly craft would be required to achieve the difficult equipment performance and schedule requirements demanded by the war. In the early days finding qualified craftsmen to hire was extremely difficult. Thus a homegrown apprentice program was required. In the latter years of the apprentice program, subject matter became less strongly connected to nuclear technology and included more of the emerging electronics industry subjects. Beginning in the late 1960's and continuing through the 1980's, craft activities migrated away from formal direction by engineering development groups and began increasingly to be directed by maintenance managers. This trend was driven by a reduction of major programs and growth of numerous smaller and technologically unrelated research activities. Although individual Instrument Technicians continued to be assigned to some I&C development activities, many returned to a central shop environment. This period found new and smaller development activities in I&C that could best use the expanded maintenance activity "as-needed" rather than full time. It was in this environment that Craft Supervisors took on a more central role as arbitrators and coaches to respond to competing manpower requirements from research groups and major ORNL operations that required regular Instrument Technician support to assure product quality and timely delivery.

In this period the I&C Division also responded to a changing funding and national energy focus by forming development groups that specialized in application of new commercial technology to Laboratory programs. Again, the need for highly specialized Instrument Technicians became apparent. At this point the apprentice program had become too expensive and narrow in its scope of technology and was terminated. The labor market was now rich with highly trained and experienced craftsmen who desired to associate with a national laboratory. Burroughs, IBM, AT&T, US military, other national science laboratories, and many more were sources of these already highly-trained new employees. Many new Instrument Technicians, as they were hired into the journeyman classification, went to work immediately in the I&C Division and non-I&C customer needs.

Several I.T.s Recall:

Richard A. Mathis Recollections

"I hired into the I&C Division of ORNL on February 3, 1964 as an Instrument Technician Apprentice. At this time, ORNL was operated by Union Carbide Corporation, Nuclear Division. I entered an ongoing apprentice program class in the middle of a program due to my past experience. I completed the apprentice program in November 1965. I left I&C in May of 1976 and transferred to the old Health Physics Division who changed their name several times before I retired at the end of December 1994.

I&C transportation was very scarce. Most I&C shops had a very unique aluminum two wheeled heavy duty cart that was designed and built in plant to move instruments between customers locations and the shops. Electronic stores was located behind building 3500 for many years and was convenient for those working in 3500. Most shops had their own stock of commonly used parts for many years. In the early years, the plant stores system was open in that customers could go back in the bins and shelves and get what they needed, fill out the appropriate IBM card and go on their way. At this time, clerks helped people find what they were looking for, stocked shelves, and processed the IBM cards. At some point in time, they went to a closed stores system where no one other than a stores clerk could go behind the counter without an escort and the clerk had to pull all items from the shelves and bins and process the check out documents in addition to their other duties. This did reduce the problem of people not accounting for the items they removed from stores, but many man hours of technician time was wasted waiting in line for items needed in the repair and fabrication process. In the 1950's and before, instrument circuits were mostly vacuum tube and analog based. Readout devices were also analog with a few electro-mechanical printers.

The transistor was new and expensive and didn't make much of an impact in the instrumentation field until the 1960's. In the 1960's, I&C started training its technicians in transistor circuitry since new instrumentation and I&C designs were now taking advantage of a wide variety of multipurpose transistors and solid state circuitry that were then affordable. Analog readout devices were being replaced with digital readouts that were first various styles of glow discharge tubes, then came the backlit numerical panels, to the seven segment displays, to the LED based displays, to the LCD displays. I&C had their own printed circuit shop and fabricated many circuit boards for specialty and I&C designed instruments. Robert Maples worked in this shop for many years. At first it was single sided boards, then double sided boards, and then manufacturers went to multilayer boards. I don't know if I&C was ever able to make multilayer boards.

My first assignment was in a process control shop supervised by Jim Day in building 3500. This shop maintained electronic and pneumatic control instruments and readouts. In this shop was a recorder repair shop which specialized in Brown and L&N recorders and controllers. Howard Frazier and Jim Knox worked in the recorder shop. I first worked with Charlie Hicks. Later I worked with Jack Richardson after Charlie left the company and went to work for ARAMCO in Saudi Arabia. Jack's specialty was pneumatic instruments of which Foxboro and Taylor were popular name brands. I transferred out of this shop at the end of June, 1964.

From July 1, 1964 to July 2, 1965, I worked in a Field Instrument Shop in building 4508. The supervisor for this shop was Phillip P. Williams. This shop worked on Metals and Ceramics Division instrumentation. Much of this work was with process controllers and laboratory instrumentation. Mickey Lewis worked full time in the Creep Lab. Bill Hicks and Jim McNeilly were technicians in this shop. They played cards at lunch time in this shop.

From July 6, 1965 to January 27, 1967, I worked in the Audio-Visual Shop in building 3500. John D. Blanton was the general supervisor with

Sanford DeHart and Jack Inman as shop supervisors. Joe Culver and, John Goans were technicians in this shop. The radio shop was a satellite shop of the Audio-Visual group with Eugene Hatfield and Roy Pollard. This shop supported ORNL meetings by operating and maintaining projection, recording, and sound systems both on and off site. Popular off site meetings were held in Oak Ridge, Knoxville, and Gatlinburg, TN. The shop also maintained laboratory pagers, two-way radios, and emergency alarm systems. On occasion, I would work in the radio shop and help Hatfield. I remember on one occasion that ORNL got a batch of Motorola two-way radios off surplus from Sandia Laboratory to upgrade the ORNL radio network. Turned out that the internal wiring in those radios had dry rotted and when they were serviced to convert them to the ORNL frequencies, we had 4th of July fireworks. The circuit shop ended up rewiring many of these radios making them expensive after all rather than free.

From January 30, 1967 to May 31, 1968, I worked in the Radiation Detection Shop in building 3500. John D. Blanton was supervisor of this shop. This shop maintained instruments for the Health Physics Division. The shop in building 3500 worked on stationary instruments and alarm systems and a satellite shop called the Calibration Shack worked on portable instruments. In this shop I worked with John Basler, George Kwiecien, Clyde Moree, Jim Payne, and Marion Dinkins. Each technician was assigned an area of the plant to maintain. Once a month, we would help each other test all alarm systems in our areas. If you were able to get all the instruments in your area in good shape and didn't have very many breakdowns, you would have some slack time during the month. We interfaced with Roland Abele's group who designed, built, and maintained special radiation detectors. They did all the maintenance on the "beer mug" alpha detectors. Clint Miller and Charlie Fowler worked in that lab. Hugh Wilson was an engineer that designed specialty radiation detection instruments and we worked with him frequently. This shop played cards at lunch. "Big Ears" Bolton from Industrial Hygiene and Don Box usually came to play cards. Once we set up an air whistle which was connected to the plant air system and was blown to signal the end of lunch break. That didn't last long since the whistle made so much noise.

At this time, I haven't found any record of where I was from June through December 1968. From January 1, 1969 to October 7, 1969, I worked in the Computer and Analyzer Shop which was located in building 2506 next to the time office. This shop later moved to building 3500. Supervisors for this shop was Earlie McDaniel and Joe A. Keathly. In this shop I worked on input and output devices with Ralph Gitgood and Don Prater. At this time, most of these devices were electro-mechanical. The popular media was punched tape, IBM cards, and various styles of printers. Teletype machines and high speed tape punches were popular. For printers, Teletype machines, line printers, dot matrix and computer controlled IBM typewriters were common. Optical high speed tape readers were used for computer input.

From October 7, 1969 to May 31, 1971, I worked in the Panel Shop in building 3500. The supervisor in this shop was John Frisbie. During this time I worked on several projects. Don Miller and myself worked for Ray Adams group of engineers wiring computer interfaces using wire wrap technology. Jim Jansen and Tom Hutton were two of those engineers. We were stationed in a utility room off of the panel shop. During this project, I went on vacation for a week or so and came back to find that my work bench, which was nice and had several drawers, had been hijacked by the division director, Cas Borkowski, to be used in his lab upstairs. In it's place was a rickety work bench with maybe two drawers on it. It was almost impossible to work on in that it moved every time you touched it. After some complaining, the work bench was reinforced which made the top steady. While in this shop, I worked on a project testing thermocouples with Norman McCullough. I also worked with Larry Basler maintaining leak detectors. In this shop I worked with David White fabricating process panel systems.

From June 1, 1971 to May 7, 1976, I worked in the Field Instrument Shop located in building 4500S. Herbert G. Lingenfelter was the supervisor. This shop maintained instrumentation primarily for the Analytical Chemistry Division and field instrumentation for the Environmental Sciences Division. This shop also supported a shop in the research arm of the Health Physics Division. Howard Barnawell worked in the Health Physics shop and was replaced with Howard Barnett when Barnawell retired. Ed Collins, Troy Chambers, Brent Davis, Ben Carpenter, Larry Lane, David Maxey, Howard Barnawell, Howard Barnett, and myself were some of the people working out of this shop at this time. Analytical Chemistry had a research project to study cigarette smoke and had some interesting smoking machines that collected the smoke for analysis and also subjected laboratory mice to the cigarette smoke to study the effects on the mice. Most of my time in this shop, I worked with Wayne Johnson and Bill Walker of the P&E Division on a miniature fast analyzer project managed by Chuck Scott of the Chemical Technology Division and funded by NASA. In the early days of this project, Lou Thacker was also involved in the design of many of these prototypes for Chuck Scott's programs. I came to this project to assist Ed Collins in the fabrication, packaging, and maintenance of these analyzer prototypes and their auxiliary instruments. These analyzers were mainly used for medical analysis but could be used for other chemical analysis. The technology developed in this program is used today in medical laboratories around the world.

Not very long after I arrived on this project, Ed Collins left the company. Later, David Maxey came to ORNL and assisted me in working on this project.

The 4500S shop soon became crowded and David and I moved to some vacant space in 4500N in the Chemistry Division area. Wayne Johnson was very instrumental in the design of these analyzers and their auxiliary components and was awarded numerous patents on this project. At least one year he was recognized as the inventor of the year for ORNL. The technicians in this shop were one of the closest groups I worked with in I&C. We went fishing and camping together, had family picnics together, and helped each other on personal projects outside of work at ORNL. At lunch time, checkers and chess was a big activity. From time to time we would have fried baloney for lunch which was a big hit. There was a black book that we recorded the many famous sayings of Larry Lane that would draw an argument on the spot.

On May 10, 1976, I transferred from Instrumentation and Controls Division to the Health Physics Division. There I went to work for Lucas G. Christophorou on a gaseous dielectrics research project. On April 1, 1984, Martin Marietta took over the contract to operate ORNL. Through the years many of us whined about how Union Carbide operated ORNL. Soon after this date, many of us would have been glad to have Union Carbide back."

Richard Mathis in 1974, was one of the first I.T.s to receive a Patent for his invention, shown here is Paul Hill giving him his letter of recognition.

Bill Koch's Recollections

Bill Koch recalls "It is my opinion that the greatest contribution of the I & C Division to the success and growth of ORNL was the embedding of engineers and technicians into the research divisions. Placing engineers and technicians into direct and near daily contact with the research scientist created a vital synergy for ORNL scientific accomplishments.

My best personal experience was at the Oak Ridge Electron Linear Accelerator (ORELA). Here several instrument technicians supported the maintenance of ORELA in addition to I & C Engineers. The Engineers both supported ORELA in addition to designing experimental apparatus needed by the ORELA physicists. A few physicist and engineers designed special detectors that were used at the Los Alamos test site where atomic warheads were tested and scientific experiments bootstrapped onto these tests to give energy data beyond the capabilities of the available accelerators. The designs were fabricated by the technicians when the accelerator was operating.

Another rewarding experience was supporting the Environmental Science Division where their scientists were measuring the effects of radiation on animals and the environment. It was strange to see a turtle strapped with duck tape to a sodium iodide detector to measure its radiation uptake. A lot of groundwater was tested by Cerenkov counting, non-coincidence liquid scintillation counting that measured the leaching of radiation from nuclear waste burial sites.

Later, I was involved in supporting the Oak Ridge Reservation Environmental Monitoring system. This system provided measurement of the gaseous effluent (stacks) of the 3 Oak Ridge plants in addition to water monitoring at White Oak Dam and the Melton Branch. Fallout monitoring was important work and a serious concern in the 80's.

While many complained of the high cost of technical support (technician's charges) I did a fabrication job with Jim Blankenship as the designer of a 10 channel summer amplifier used at the CERN. I fabricated and he tested the 22 modules before shipping these for an experiment at the CERN. The CERN wrote back that of the 220 detector channels, only one did not work on turn on, they were amazed and pleased with this quality work. This job was done at a lower cost than a bid received from a local fabricator."

The Seeds of Maintenance Management

It was in this environment that I&C management saw the need for a separate management organization totally dedicated to hiring, training, work assignment and union conflict management. In the rush to respond to early program needs and later to development group specialization, management had accumulated a large backlog of union grievances. It was during the 1950 to 1970 time frame that the Atomic Trades and Labor Council, a bargaining unit for all trades in Oak Ridge and a consortium of 16 A.F. of L. Craft Unions, became effective and active in pressing for better benefits for its constituents. The ultimate union structure was the result of an election held by the Department of Labor at the X-10 site. A byproduct of that effort was the empowerment of each craft union group to become active in monitoring the detailed day-to-day activities of engineering groups and Instrument Technicians in shops.

For the uninitiated, Instrument Technicians were the best-educated and most creative group among the various crafts. They were required to understand a broad range of technological, scientific and fabrication skills to be effective. This fact is well-illustrated by A Reactor Division report of the time that list tasks that would have been performed by I&C ITs:

"Graphite Reactor Annual Report for 1957 – excerpt, J. A. Cox, W. R. Casto Control System Upgrade:

The control system for the Graphite Reactor was modernized during CY 1957. Fission chamber and ion chamber channels, both with period and log indication, were installed, in order to make room for these new instruments on the main instrument panel, the inlet and outlet temperature recorders were moved. At this time, part of the old recorders installed in 1943 were replaced with modern instruments. It was necessary to use the reactor instrument hole formerly occupied by the No. 1 galvanometer chamber for the ion chamber. For this reason, only the No. 2 galvanometer is now in service. Scanner hole No. 5 was used for the fission chamber.

Standard annunciator units of the plug-in type I have been installed to replace the original annunciation equipment. While this work was in progress, it became evident that the original 24-v safety circuits contained relays that were in need of replacement. To complete the standardization of the reactor controls, it was decided to replace the old system with 110-v equipment. By the end of the year this was practically complete.

All the chamber power supplies have been placed in suitable racks in the control room. At the end of the year the Reactor Controls Department was revamping the wiring and developing the information necessary for a new and authentic wiring diagram for the Graphite Reactor."

Central portion of Control Panel of Graphite Reactor

Although complex to describe to the uninitiated, a simple example will demonstrate the point to be made. A hypothetical reactor control room supervisor has requested that I&C replace a metal panel that currently contains a 1950s-vintage temperature recorder capable of continuously recording two temperatures from the reactor vessel. The goal is to install the current technology twenty-channel temperature recorder and include a pneumatic transducer and heavy duty A/C power relay. In order to produce the most professional look and best quality control of the new panel, Instrument Technicians are required to punch holes in the panel, mount equipment and test the functions of all the devices before delivery to the reactor for installation. Thus they have assured 100% likelihood of no rework because of discovered flaws after installation. The "catch," union craft-wise, is that the pipe fitter at the reactor may choose to take issue with the IT performing "his work" by installing and or testing the pneumatic transducer (it has plumbing). Likewise the electrician assigned to the reactor might well take issue with the IT installing and or testing the power relay. This one, relatively simple job, has now produced two grievances. In union parlance, two grievances require two craft supervisors and two union representatives to meet and debate the perceived crossover of craft responsibilities with the goal of reducing future such discussions. During this time frame, the grievance list had grown from a few per year to several hundred unresolved by 1979. To this writer, the two primary reasons for the establishment of the Maintenance Management Department were programmatic and union/management challenges.

Maintenance Management Department Established July 17,1978

Paul and Muriel Hill. Paul was selected to head the Department.

Division Director Bill Eads Said:

Selection of a suitable individual to lead this new Department was relatively easy after listing all requirements, (1) foster a team spirit with the newly empowered maintenance team, and (2) deal effectively with the grievance backlog. Paul Hill was selected from within the organization because of his easygoing nature and natural leadership qualities. Paul immediately set about meeting with union stewards and shop foremen to develop a better dialog. Within a year or so the remaining backlog of grievances was reduced to fewer than a dozen.

A key point of contention was loss of the Red Line classification of Instrument Technician. Apparently, Engineering Section Leaders who took the lead in earlier union negotiations were unfamiliar with the original purpose of the higher pay category and thus did not defend it in the face of a heavy assault by the Electrician Craft union representatives and the ATLC. The result was a loss of bargaining power for ORNL to hire the best and brightest candidates for the IT craft.

The next few years, 1979 to 1984, were characterized by a relatively tranquil union and company relationship. Staffing remained at a high level in I&C maintenance. The tasks required from ORNL customers were well funded and diverse, supporting the continuation of internal instrument manufacturing, quality control and procedure writing. ORNL program managers who were paying for these services sometimes took a central role in documentation and operation manuals needed to accompany these instruments. Health Physics Division and Reactor Division were prime examples.

Associated with the skills needed to support Reactor Division was the Hot Cell operations. ORNL's role in DOE programs involved quite a few different radiation monitoring and radiation safety monitoring instruments as well, in the late 1980s. A large fraction of these devices were original I&C Division designs and were, consequently, most efficiently supported by I&C IT's. Some of the original vacuum tube devices designed during the pre-transistor era were eventually replaced by a second generation of ORNL-designed equipment. Development of nuclear reactors was a major part of ORNL work between 1950 and 1970. Designs with names such as ANR and BSR, LITR, ORR, MSRE required a major emphasis on design and construction of fission chamber devices and their associated amplifier circuits suitable for use in each of these uniquely different reactor configurations. Often these detector chambers and their circuits were arranged as one package built by Instrument Technicians. Early on they had miniature vacuum tubes embedded in a snake-like assembly, associated with the detector. Later transistors replaced the vacuum tubes, as shown in the 1962 photo, below.

A key example of this transition began in the mid sixties with the second generation of nuclear reactor controls. ITs selected to work on the fabrication and installation of these devices were carefully selected from across the various I&C shops. Installation and initial testing of these new devices drew the

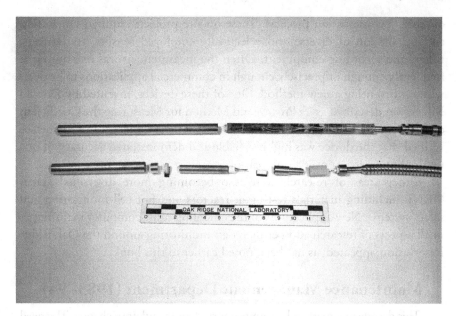

This snake-like assembly was a fission chamber and pre-amplifier which could be fed through a conduit into the center of a reactor core (for reactor startup), then withdrawn gradually, as the reactor power increased. It was designed by I&C Engineer Dominic Roux and was fabricated by I&C Instrument Technicians.

attention of Department of Energy regulators and ORNL nuclear engineers from I&C, Physics, and Metals & Ceramics Divisions. This closely monitored activity drew the I&C team closer in teamwork and eventually became a model of effectiveness for safe operation of a nuclear research reactor.

February 1, 1977 was the first of several Presidential Fireside TV Chats held by Jimmy Carter (re-initiated from the 1930's as radio chats by President Franklin D. Roosevelt) focused heavily on Energy. In this talk, President Carter stated unequivocally that America faced an energy problem which called for a long-range comprehensive plan and introduced his Energy Policy Initiative which focused on conservation. In the years closely following this "Fireside Chat," ORNL took a central role in reacting to this challenge. The Energy Division was created, and I&C became a strong partner in the research. At one time almost 10% of the labor dollars coming into I&C were energy research related. Some I&C engineers, weekly and hourly were assigned to or on loan to the Energy Division. Projects numbered in the hundreds and extended to military facilities across the US. In the case of the Army, the Corps of Engineers were mandated to design and install energy saving devices at all installations. By this time Don Miller had been promoted to monthly and was given the job of group leader over six engineers who were employed full time on energy conservation projects. These on-site projects required fabrication and installation of diverse and eclectically employed sensors, transducers, timers and recording equipment. Often the measurement was one not done frequently enough or precisely enough in commercial applications to be done without inventing a new method. One of these devices, invented by Donald Miller, was described as "A Process and Method for Measuring the Coefficient of Performance of a Heat Pump". Field tests at the University of Tennessee verified that the device was highly reliable and demonstrated accuracy of one tenth of one percent over a one year operation.

Other areas of research were also becoming more diversified across ORNL including materials testing at reactors and hot cell monitoring and safety systems. New applications of the emerging mini-computer mainframes in all aspects of research and perimeter air monitoring around the Oak Ridge Reservation appeared, as has been noted earlier in this book.

Maintenance Management Department (1983-'94)

This decade was marked by significant staff and regulatory change. The need for Paul Hill's management style with emphasis on labor relations was ending and Don Miller was being positioned to be his replacement. The labor relations management emphasis was changing to a DOE-driven nuclear safety re-emphasis and introduction of refined and complex procedures for maintenance activities. Laboratory Directors and line managers were participating in strategic and detailed planning to meet the new challenges. I&C Division's maintenance department was seen as an important link in the newly emerging company response to expanding regulatory requirements and closer external monitoring.

It was with this background that Paul Hill was to have retired in the fall of 1985. Don Miller recalls that he had been reassigned to MMD in June of that year to serve as a General Supervisor over three major I&C shops; Process Instruments and Panel Fabrication, Radiation Monitoring, and Reactor Systems. His prior experience, while hourly, in the Molten Salt, Oak Ridge Research, Graphite and High Flux Isotope Reactors became invaluable in this new assignment. Twenty-nine highly experienced hourly and 4 weekly payroll Instrument Technologists were assigned to this area of work. C. G. Allen and J. D. Blanton were the craft supervisors in the areas.

Miller found the support work effective and adequately staffed; however, I&C Division engineering groups were not strongly integrated into day-to-day tasks as had been the case in earlier years. His initial tasks were to acclimate himself to the new, for him, organizational structure and begin to draw on traditional I&C engineering resources to bolster "maintenance work" where it amounted to redesign or technological update. As the date for

Paul Hill's retirement drew near, it was apparent that his replacement would be faced with numerous challenges. The previous decade had been strongly focused on labor relations and I&C instrument shop diversification. During this same decade, the incessant march of technology had left the maintenance staff and its tools in an increasingly obsolete posture. Few I.T.s had training in integrated circuits, calibration and standards practice, procedure writing or electric signal communication standards.

Untimely Death & New Appointment

September 6, 1985 was a significant date for the Maintenance Department. Paul Hill passed away unexpectedly three weeks before his official retirement date. The Division leadership had to choose and announce his replacement a little earlier than previously expected.

Bill Eads division director, remembers:

"On April 1, 1984 with the assignment of Martin Marietta as operating Contractor of ORNL, (later Lockheed Martin), I was appointed Division Director of the Instrumentation and Controls Division. The division was beginning to move into Work-for-Others to some degree and that took off dramatically during my tenure.

Just a few months after being appointed, I remember struggling with the decision as to who would replace Paul Hill as head of the I&C Maintenance Management Department. I actually found it a relatively easy decision to make in terms of the best person for the job. The challenge was to figure out how to do it without alienating the 'old guard' folks in Maintenance. D. R. Miller was the person selected to take up the task."

During the previous three months Don had completed four major tasks of interest to both the Department and the Division in his position as General Supervisor.

C. Defined and initiated a Division maintenance policy for Programmable Logic Controllers (PLC), which had recently emerged as a primary choice to replace traditional electromechanical relays in control system design.

CI. Arranged for two I.T.s to attend a factory PLC maintenance course. Also pursued personal conversations with craft supervisors, field engineers, and customers of I&C about the need for more formality in I&C's support of PLCs.

Programmed Logic Controllers were specified by I&C Division engineers, often without adequate training for the I.T.s.

CII. Reviewed the then primary record keeping system for Division maintenance of ORNL's instruments and controls. Developed and documented an upgrade plan for Maintenance, Accountability, Inventory System (MAINS). The MAINS record keeping system had been devised by Joe Keathley and Karl West in the 1970's to consolidate several disparate record systems then recorded with pencil and paper or typewriter. MAINS was written in FORTRAN and data was entered through the use of punched IBM cards and the system used a computer in 4500 building. Essentially, calibration data, repair data, installation/removal and location of equipment was tracked by the system. Individual shops or customers could request monthly summaries of work performed or scheduled to be done from the ORNL central computer staff. Craft supervisors would prepare IBM card sets to match the planned work and provide them to individual I.T.s on a daily basis for completion. It was clear that mini-computers, now small and cost effective enough to be available in 3500 building, could be used to process maintenance records. Job data could now be entered directly into the computer eliminating the punched card intermediate step in data storage. In addition it was possible to include timekeeping information for each job. The Measurement and Controls Section had incorporated mini-computers in their design work and in time keeping for division staff. Through a mix of MMD staff, summer undergraduate staff and Division staff, the Maintenance, Accountability, Jobs, and Inventory Control Program (MAJIC) was developed.

CIII Held formal and informal discussions with ORNL's Central Engineering Division with the purpose of strengthening cooperation and use of resources.

Sandy Odom Remembers:

Editor's Note: Sandy Odom was one of two people who were diverted from the clerical pathway after doing exceptionally well in a "programmer's aptitude test" that was administered in Charlie Mossman's Section to several I&C Division clerical people (the other person who did well was Jane Cordts). Sandy then worked as a computer programmer's assistant to Roy Simpson, in the I&C Process Control and Instrumentation Section. Sandy is now System Administrator of the MIDAS/Facility Service Center in the UT-Battelle Facilities & Operations Directorate.

Evolution of MIDAS from the IBM Card System.

Roy Simpson was an integral part of the I&C maintenance process. Initially the maintenance history and equipment information was tracked totally on a card system. At some point during the 1970s, Roy adapted that hardcopy card system to an IBM driven punch card system that was written in Fortran Programming Code. This managed the equipment and maintenance Preventive Maintenance Schedules for that equipment. Each week, Roy would take the two boxes of IBM Cards to the computer lab. In these cards (which were several hundred per box), the program code and the equipment data cards would be processed to generate the output listing of the next due PMs. This information processed on a IBM main frame system in the Computer Center in 4500N.

This process stayed in place into the early 1980s to around 1985 when computer technology had advanced to the point that there was a new method they wanted to migrate to. The microvax computer era had evolved that had a database called 1032. This was a later evolution of the 1022 main frame database system that came along in the early 1980s. Management determined, with Roy's guidance, that this might be a valid migration path for the old IBM card system. Roy Simpson worked with Terry Stansberry from the maintenance organization to explore the use of this "new" technology to replace the IBM Punch Card system.

Terry worked this with Charlie Barringer, the internal MMD programmer. They developed the database beginnings for the MIDAS system and migrated the business information from the card system to the first MAJIC database on a microvax computer located in Building 3500. This included Equipment Inventory, Preventative Maintenance Schedule, Work Order, and Calibration records and Traceability Databases. Charlie and Terry designed the first work order process for tracking the PM work as well as other repair work details. This new program replaced the IBM Punch Card System in October 1986.

In 1989, Charlie Barringer retired and Terry Stansberry turned the system over to two summer student interns who worked to build an umbrella structure for the Maintenance Program called Maintenance Information and Data Acquisition System (MIDAS). MIDAS was to include MAJIC which is the Maintenance Accountability and Job Inventory Control system. In addition, the MIDAS system contained a personnel database, a training database so technical training could be documented and tracked, a document management system for procedures, and other business processes for managing maintenance. This was the beginning of the I&C Maintenance Management computer-based business management system. The two summer interns left to return to college and the system was turned over to Sandy Odom who managed it through several upgrades and evolutions over the years. Sometime around 1994 the MICRO VAX was replaced with a VAX 6310 computer system and the programs were migrated over from the MICRO VAX.

This MIDAS system was the core business process until the I&C Division was dissolved. After the division was dissolved, the MIDAS system work control functions (including MAJIC, Documents, Personnel) were maintained until November 2006 when the Work Control system was migrated to a new ORNL System called Facility Service Center (FSC) that is the repository for all ORNL maintenance activiites. Sandy Odom continues to support the MIDAS sytem which is on a VAX computer. This VAX Server moved from Buiilding 3500 to Building 2033 and then on to Building 6310 during the time from 1994 to present.

Currently the MIDAS system is available for equipment history from October 1986 through December 2006. This archive instrument history information will be migrated to an ORACLE platform within a couple of years, merging with the FAMMIS Facility System history data since both systems are now operating in the Facility Service Center (FSC).

PicORNL02857-90

Don Miller was appointed Head of the I&C Maintenance Management Department

Don Miller's Recollections

Donald Miller wrote that his job responsibilities in 1985-86 were: Department operation, policy, personnel actions, record keeping systems, facilities and external communications. The Department consisted of 91 hourly I.T.s, 25 weekly Technologists, 10 Craft Supervisors and 2 general foremen. The Department had recently accepted 9 hourly transfers from K-25 to replace openings created by that plant's voluntary reductions in work force (VRIF). The new staff were assigned in the then growth areas; Computer maintenance and Process/Environmental support.

Don met with both maintenance and Division staff to assess major challenges before the Department. A list of 5 focus areas was developed for immediate action.

1. Review Instrument Technician training activities and plan emphasis

changes

2. Initiate regular Craft Supervisor meetings to review problems and develop guidance for solutions

3. Review the Department Management Plan to adjust for current needs

4. Replace existing records database with a more up to date unified approach written in an efficient modern data base language (PL1032). "**1032 Database**" database management system and application development environment designed to support the OpenVMS user community had recently become available.

5. Review all personnel assignments for effectiveness and potential growth

System-Wide Appointments Added

Due to the broad interaction of MMD staff with all ORNL Divisions, programs and nuclear facilities, Don Miller was assigned to several Laboratory committees in that first year as Department head of MMD. He was appointed by the Laboratory management to the "Director's Radioactive Operations Safety Committee", Lockheed Martin's "Five Plant Nonexempt Salary Review Committee", and the "Laboratory's ADPE Committee".

A February 5, 1986, Internal Correspondence of Martin Marietta Energy Systems, Inc. memo assigned D. R. Miller to the ORNL Radioactive Operations Committee (ROC).

The memo noted:

"I am writing to request that you serve as a member of the ORNL Radioactive Operations Committee to fill a position being vacated by L. D. Bates, who has served conscientiously since 1981. Your appointment will become effective immediately unless there is some reason you cannot serve.

This position is a very important part of the Laboratory's safety program to ensure that our radioactive facilities and operations continue to be operated safety. It is through the efforts of conscientious and experienced employees like you that we have been able to achieve and maintain our excellent operational safety record.

B. G. Eads has been contacted in regard to your appointment and understands that this will be an important and integral part of your work assignment. Should you have any questions about operations of the ROC, you may contact the ROC Chairman, Joe Setaro (4-5903), the Office of Operational Safety Director, Howard Burger (4-4339), or Office of Operational Safety staff member, John Alexander (4-4340).

Signed by

Ray S. Wiltshire, 4500 N. (4-8447)"

It was apparent to some of ORNL staff that a massive political change was occurring within DOE and its funding political base. Nuclear safety was viewed in an entirely different light as World War II managers and congressmen retired and were replaced by those with a less patriotic and more cautious view of nuclear safety. Over the next decade this change would become more apparent to everyone at ORNL. Often I&C would be at the traction stage of the change process, being involved in writing maintenance procedures, performing those procedures and defending solutions sometimes difficult to explain to non-scientific inspectors.

To understand this period of operations better within I&C, it is useful to review some of the comments made by Don Trauger (ORNL Associate Director) in his article published in the "Lab News", May 1984.

"How is ORNL organized, and why?" ORNL is basically organized by divisions, sections, and groups or departments. Some broader programs, however, use the resources of several divisions and may even use resources outside the Laboratory. Programs such as these are an important source of business for ORNL; in fact, well over half of the Laboratory's funding is distributed through programs. In our program organizations, division directors and their staffs are held accountable for the technical excellence of work done in a particular division, for maintaining and developing staff, and for maintaining a position of leadership in state-of-the-art research. The divisions are also accountable for job stability and suitable working places for staff members. The program director, on the other hand, is responsible for all relationships with our customers, be they assistant secretaries, office directors, or division directors of DOE or their counterparts in the Nuclear Regulatory Commission or other government agencies. But satisfying the customer is not enough. We must also maintain the integrity and credibility of ORNL. An important feature of the programmatic organization is that staff members can move easily from a completed task to a new project. Staff members can readily change programs while maintaining a stable position within the division structure and benefiting from association with colleagues within their technical discipline."

In contrast with the Trauger overview, the I&C maintenance function had been strongly influenced by local union activities and craft job lines-of-work assignment. Union steward and craft supervisor grievance resolution history ultimately drew lines between Electricians, Pipe Fitters, Sheet Metal workers and Instrument Technicians (I.T.). A continuing effort of the Atomic Trades and Labor Council was to combine the I.T. and Electrician seniority list. The successful achievement of that goal would reduce management's ability to make effective use of I.T.'s unique training and experience while reducing the exposure of Electricians to layoff in times of budget reduction.

To give the reader an insider view of this conflict, the response to a 1989 union proposal as written by the MMD department head is given.

AN ANALYSIS OF THE IBEW PROPOSAL FOR COMBINED SENIORITY LIST

The subject of combining the two seniority groups within the ORNL Electrical Crafts has been suggested at almost every major contract renewal since 1964. As we understand it, the essence of the IBEW claim is that the company would benefit from fewer jurisdictional disputes. Reading over the various claims and counter claims from previous such discussions; we find plausible reasons for taking either side of this argument. The primary fault with this traditional analysis is that it assumes something that is not provably true today. It assumes that Electricians and Instrument Technicians are of equal value to the company. It also assumes that the educational level prior to employment and after employment remains equal. Each year which passes finds this to be less true.

We believe that it is important to consider some of the reasons that the Union would favor such an arrangement. The Instrument Technician is not a recognized craft within the IBEW. In affirmation of this fact the Union has, at times, eliminated local representation by the Instrument Technician Steward in the daily business activities (no vote even though they represent the largest craft group at ORNL). In 1961, the union local had a major influence on the company decision to eliminate a premium pay classification for some Instrument
• *Technicians. During the same period the Union was able to gain control over jurisdictional disputes because the company was disposed to minimize conflict. In those days the major jurisdictional battles were over the mechanical activities of Instrument Technicians such as panel work. In the face of such open hostility, the Instrument Technicians are not pro Union. A combined seniority list, would greatly hurt morale, place Instrument Technicians in an untenable position and hurt recruiting and work assignment. The company now has an opportunity to regain control of its own destiny in a national political climate that is more moderate toward the Company/Union relationship.*

In addition to the historic conflict, it seems likely that a new set of disagreements would arise following the merger of the seniority list. Based on long years of experience with the IBEW, we can expect a demand for equal training, freedom, overtime, promotion opportunities and quality of workspace now enjoyed by the Instrument Technician Group. Once made, this decision would be essentially permanent. The Laboratory should carefully consider the potential cost of the proposed merger. The Laboratory has unique facilities and research efforts that require state of the art instrumentation and close coupling between the engineering design effort and the follow on maintenance and calibration.

It is interesting to note that during the past ten years (1979-89), approximately 35 Instrument Technicians were promoted to weekly Technologists and some of them were later promoted to monthly engineers. At entry, the current average educational level of the I & C hourly is two years of college. The Maintenance Department annually spends about $150,000 on technical training for Instrument Technicians and Technologists. The Division Director recently said "Having the Instrument Technicians as part of the I & C team is an important part of our ability to attract high tech work". The scientific equipment and industrial controls industry is producing new equipment which flows into the Laboratory each year and which must be supported. This equipment is more complex each year and is increasingly being interconnected by networks and distributed control arrays. It is imperative that we continually upgrade the hiring requirements and employee training level. In this connection, four of the I & C Maintenance Technicians have achieved a bachelors degree within the past two years. Forty Instrument Technicians have an Associate Degree and eight are pursuing a BS Degree at present. The recent formal signing of theDOE Accreditation Order will require the Company to change the 90 day probation period to 18O or 36O days such as TVA has done for years. It will, also sharpen the lines of differentiation between the Electrician and the Instrument Technician."

1990 ATLC/MMES CONTRACT NEGOTIATIONS
SUBJECT: Proposal to Combine the Electrician & Instrument Technician Crafts Made by the IBEW.

This subject has its origin in several past actions on the part of the Contractor and the Union. It has been reported that the original organization of the labor pool at ORNL was done by AEC to demonstrate that nuclear reactors could be supported by a unionized work force.

As a result of the original policy and a labor agreement on October 31, 1947, the Laboratory policy has been to recognize that the requirements for the Instrument Craft "were different from the majority of the industrial instrument groups, but was necessary due to the conditions prevailing". That is to say, R&D can only be supported by highly trained Technicians who were hired for that purpose. On January 18, 1954, Local 760, IBEW reached an agreement concerning the work assignments between the two groups at the X-10 Plant. This agreement delineated the assignment of work with respect to cabinet fabrication, power distribution, controls and maintenance of components. On September 1, 1964, the Company offered a trial suspension of the jurisdiction between the Electrician and the Instrument Technician classifications. This was to be a trial period to evaluate the efficiency and problems that developed. Messrs. Borkowski, Seagren and Lieber signed the stipulation.

On November 4, 1974, a report was issued by Strohecker and Briggs, Y-12 staff members, giving an independent view of the merger issue. This study was done entirely from the perspective of management at the suggestion of ORNL management. Several interesting opinions and facts emerged from this study. They note that I&C was formed in 1953 by combining the instrument research, development, and design groups of several research divisions with the design and maintenance groups of the Engineering and Maintenance Division. This was done to provide a greater central capability for doing instrumentation research and development. They note that historically the majority of grievances filed by Technicians are against the actions of monthly or weekly personnel. It was suggested that merger is not likely to produce a large improvement in efficiency. In summary, the recommendation was; "we find no strong incentives to merge the seniority groups, nor do we see that serious problems need result from such a merger.

In retrospect, it seems clear that the inclusion of the Instrument Technician work in the Union Contract was a serious error. No other National Laboratory has this situation; per Frank Manning - 1990.

It was pointed out by Frank Manning on April 11, 1974, that "*with the relaxation of jurisdictional lines, the two crafts will begin to diffuse into the gray area work across the entire electrical work force. The control of the craftsmen in isolated areas, essentially supervised by research workers, could conceivably result in a disastrous situation for either equipment or personnel*".

Frank further noted, "*There is no point in giving the IBEW the prerogative of directing the technological approach to Laboratory problems in an area where they have little competence*".

Signed by

D. R. Miller, I&C Division MMD Department Head."

Joe Keithley was a long serving member of the I&C Maintenance Staff.

During the 1980's two general supervisors dominated the day to day I&C maintenance planning and operations.

Joe Keithley and Carl Kunselman were solid, experienced General Supervisors. Their recommendations for assignment and promotion were generally accepted by Miller without lengthy debates. Joe had a solid career in computerized analyzers and mini computers and more recently teamed with Reactor Controls Section staff to gain control of the nuclear support information through development of a new and unique Fortran based system. Carl brought his extensive field experience, computer and process background to the task.

Carl Kunselman started out as an Instrument Technician in the Computer and Pulse Techniques Group. He was promoted to General Supervisor in the MMD department.

In a 1990 position description for the Department Head it was said that "The Department must always strike a balance between adherence to the labor contract and the need to be innovative and progressive in a dynamic technical/political environment."

Throughout the decade of the late 80s and early 90s ORNL looked to I&C for significant leadership and technical capability to assist research divisions and in particular nuclear facilities for Nuclear Regulatory Commission compliance and safety. Revised and new DOE orders had the weight of federal law insofar as the details of facility operation was concerned. Regulators came from Washington and other federal facilities to visit on a regular basis to inspect records and written procedures performed by MMD staff. Each written and peer reviewed maintenance procedure was subject to close scrutiny by on-site customer staff and these regulators. Occasionally failures of compliance with similar procedures at other federal facilities would energize the Feds to come and put ORNL level of compliance under the microscope. Leading our ranks in MMD to defend and react to these audits was the Process Instrument Maintenance Group and the HFIR shop. The PIM group of hourly, weekly and monthly staff grew out of earlier operational requirements well described by Bob Effler in this section. A regular internal progress report described their activities as follows:

Process Instrument Maintenance
(R. A. Vines, M. E. Boren, J. P. Jones, E. P. Trowbridge, and D. G. Raby)

The Process Instrument Maintenance Group includes the main shop in Building 3500 and three satellite shops in buildings 2519, 3005, and 3606. The group is staffed by 1 supervisor, 4 engineering technologists, and 11 instrument technicians and is primarily responsible for providing process instrument systems maintenance, calibration, and fabrication throughout ORNL.

The main shop provides field maintenance and calibration of all customer instruments as required; calibration of secondary field standards; fabrication and testing of thermocouples; inspection and calibration of new instruments; and design, fabrication, and installation of control loops and associated electronics. This shop is also responsible for the maintenance and calibration of leak detectors and vacuum systems. One technician in Building 2519 is responsible for the maintenance and calibration of all steam plant, waste water treatment plant, and laboratory water supply system instrumentation.

Environmental and Radiation Monitoring Maintenance
(B. L. Carpenter, D. M. Duncan, R E. Gallaher, W. R Blodgett, and W.E. Wright)

The E&RMM Group includes 2 supervisors, 3 engineering technologists, and 20 instrument technicians who provide maintenance and other support for portable radiation survey, fixed-station, and other monitoring equipment for the ORNL Environmental & Health Protection Division. In addition to servicing all ORNL portable radiation survey instrumentation, two technicians also service portable units for the Radiological Survey Activities Group of the ORNL Health and Safety Research Division as well as all portable radiation survey instruments at the Oak Ridge K-25 Site. A service group of nine technicians has maintenance responsibility for stationary monitoring instrumentation; fallout; and local, perimeter, and remote air monitoring systems as well as all facility radiation, contamination, and similar alarm systems. In addition to normal maintenance and routine system checks, this group provides support in checking the new perimeter air monitoring system, a water quality monitoring system, and stack monitoring systems. One technologist and six technicians provide service to the Gaseous and Liquid

Waste Disposal Station, including normal routine maintenance and system checks. The group also provided support for checking and the successful operation of the process waste monitoring equipment and supported the upgrade and modification of the Waste Operations Control Center.

Special Electronics Maintenance
(K. L. Allison, J. S. Riggs, and H.E. Smith)

The Special Electronics Maintenance Shop consists of one supervisor, two engineering technologists, and six instrument technicians. To provide efficient maintenance support, the shop is divided into three sections.

1. The Analyzer Maintenance Section provides maintenance support on pulse-height analyzer systems, liquid scintillation counting equipment, phototypesetting equipment and miscellaneous nuclear instruments located throughout ORNL.

2. The Test Equipment and Oscilloscope Calibration Section provides maintenance and calibration on test equipment and oscilloscopes. This section is staffed by two instrument technicians and one engineering technologist."

In the early 1990's DOE became more interested in contractor infrastructure devoted to process improvement related to maintenance activities. A new Maintenance Improvement Plan (MIP) was transmitted to all contractors for implementation.

Maintenance Management Department Renamed Technical Support Department

I&Cs Maintenance Management Department (MMD) was to become the Technical Support Department (TSD.) TSD became involved early in the re-sizing and refocus process by seeking volunteer hourly staff to study the plan and make suggestions for process improvement. In a memo to TSD supervisors D. R. Miller said: "The purpose of this memo is to make you aware of the key issues noted for corrective action, and what we propose to do about them. As you review them, please make note of any item that you, personally, can help facilitate to a successful conclusion.

RECOMMENDATION A: - The Department will revise its Oversight Committee Charter to include evaluation of each element using a graded approach, and to include all staff involved with the Maintenance Improvement Plan. Expected conclusion of the effort is April 22, 1994. *(Graded approach was terminology suggested by DOE to consider risk versus cost in solutions to maintenance challenges.)*

RECOMMENDATION B: - The existing job check-list will be reviewed for completeness and made available to all shops by February 28, 1994. Each supervisor or task leader will be expected to include the check-list items in all pre-job planning requiring it. *(this check list was devised to insure that the job completion process proceeded with due attention to clear instructions from the customer, complete documentation of the calibration standards required, employee numbers of those performing the work and hours charged to the customer account)*

RECOMMENDATION C: - The P&E Division will develop criteria for vehicle utilization reviews and possible redistribution by September 30, 1994. *(A continuing problem for I&C staff was the limited number of vehicles available to workers. Jobs requiring a quick turn-around and located some distance from building 3500 were frequently delayed due to non-availability of vehicles. The problem was partially solved by Department /Head Miller when he bypassed the DOE policy and purchased golf carts with overhead money from the department for use by ITs.)*

RECOMMENDATION D: - The Department will conduct a review of bench stock requirements, particularly focused on safety related instrumentation; to be completed by June 30, 1994. *(Bench stock had become a focus of Laboratory Management because it represented a large financial commitment to maintain parts outside Electronic Stores)*

Note that these descriptions are abbreviated for ease of review. If you desire further details please contact A. J. Millet."

This example is one of many activities and processes introduced or revised in the decade of the 90's. At that time Allen Millet was the task leader for compliance and safety issues in the Department. Allen was one of the small group of former hourly workers who progressed through higher education and "sweat equity" to weekly and then monthly grades, finally being valuable team leaders in the increasingly challenging environment of external DOE and internal ORNL pressure to be everything to everyone in the race to have zero defects, efficient union workers and reduced staffing.

A.J.Millet was a healthy contributor to the I&C Division's TSD (formerly MMD) Department..

During this externally driven struggle to understand and react quickly to changing requirements, the department found itself again without internal division allies in the instrument and digital controls engineering needs of ORNL. Bob Effler had traveled a path similar to that of Millet. He had recently been involved in department improvement projects and staff training. While foraging for a focal point to achieve department capability in what we referred to as Field Engineering he was selected to lead a new TSD Field Engineering Group whose sole task was to handle small internal-to-ORNL design tasks for other ORNL Divisions.

Bob Effler (shown here at his retirement) was one of the managers who helped Don Miller and the Department managers who succeeded him in the I&C Maintenance Management Department and the Technical Support Department. Bob later became manager of the Metrology (Standards) Laboratory, when it was Separated from the I&C Division.

Initially he pulled together 9 TSD staff engineers who were previously assigned to individual shops and created the "Research Support Group" (RSG.) This synergy and focus quickly became a success story. Some limited access to the I&C engineering groups was still available for consultation, but 95% of the work proceeded with Effler's staff who were all originally hired as hourly workers. By the mid 1990s the capability of these staff engineers was such that work outside ORNL came their way. One such job was for design work to be installed at the nuclear reactor at University of Michigan. This was an, at-the-time, ground-breaking data acquisition and control system (DAS) that allowed researchers to conduct experiments to determine the rate of embrittlement caused by a neutron flux to different steel alloys intended for future reactor vessel designs. Because, at the time, the only reactors at ORNL were unavailable, RSG engineers developed this remotely-controlled DAS so that ORNL researchers could conduct these experiments, physically located in Michigan, from their labs at ORNL.

Other notable RSG successes were:

- A monitoring system to detect and characterize leaks in barrels of radioactive materials (^{137}Cs, ^{60}Co, and ^{90}Sr) submerged in an underground water-filled canal.

- The first "spectralizer" interface between a Multi-channel analyzer and a liquid-scintillation spectrometer.

- New digital control systems for the Materials Irradiation Facilities at ORNL's High-Flux-Isotope-Reactor.

- Computer-based open-channel flow monitors for all of the groundwater streams flowing through ORNL property.

- A "ramping-titrator" to eliminate data-stream bandwidth saturation in an isothermal calorimeter (This resulted in a patent disclosure.)

- A system to analyze the level of radioactive contamination in underground water on the island of Kwajalein.

TSD's Research Support Group staff filled an important niche in the needs of the ORNL research organizations for practical, "hands-on" design engineering at a time when the previously available I&C engineering resources were being diverted away from supporting other research programs to performing research projects of their own.

Bob Effler's Recollections

"I was hired by I&C Division in May of 1979 as part of the great "Burroughs influx." Burroughs Corporation, a computer manufacturer and somewhat quixotic competitor to IBM, retained a staff of "Field Engineers," assigned to maintain their customers' computer equipment under maintenance contracts. Although small by IBM standards, Burroughs did have a reputation for good quality equipment and services. The quality of service was largely the result of a corporate policy of thoroughly training their technical staff. I spent almost 25% of my time at Burroughs in training centers (We called them "stalags.") at Syracuse and Philadelphia. Much better-paying jobs and more gratifying jobs were available at ORNL, however; so, one by one, Burroughs lost most of its cadre of Field Engineers, who found it more desirable to work as hourly Instrument Technicians for ORNL's I&C Division than to maintain the dubious prestige of being called an "engineer" and trying to make a living by driving about the country fixing computers.

Among the "Burroughs" names that come to mind in addition to my own are: Lawrence Finchum, John Roach, Charlie Tompkins, Ken Pate, Daryl Valentine, Bert Harper, Dannie Sluss, Ken Wright, Ron Maples, Doug Smelcer, and Bill Gorman.

I was originally assigned to the "Analyzer Shop," a group in the I&C Division's Maintenance Management Department. This group was supervised by the irascible, "Ragin' Cajun," Al Millet. We were primarily tasked with maintaining multi-channel analyzers and liquid scintillation spectrometers. We also took care of a monstrous photo-digitizer called a "spiral reader," for a high energy physicist, and a photo-typesetter for ORNL's publications organization.

The often overlooked, but vital importance of the I&C Division's Instrument Technicians' role in supporting R&D was made plain to me very shortly after I began to work at ORNL. One of our clients in Chemistry Division had been working for years on a technique to use liquid scintillation equipment to analyze samples of solutions of alpha-emitting isotopes in the presence of a high beta/gamma background. The apparatus that he had devised, with I&C assistance, consisted of an entire "Bud-rack" of individual NIM-standard modules with a morass of coiled coaxial cables to provide some necessarily delicate fine-tuning of the timing of the pulses generated by the liquid scintillation detector. It barely worked and had to be "tweaked" constantly. One Mr. Don Prater was the I&C "expert" on this. I was assigned to help with this "monster." One day the researcher brought in a home-made NIM-type module that had been designed and built by a predecessor. It was intended to perform the same function as our entire apparatus but had never actually worked. He asked if we could take a look at it. Mr. Prater "pooh-poohed" the whole thing; but, as a courtesy to the customer, asked me to "give it a shot." It wasn't difficult to discover that the designer/builder had simply neglected to provide a simple "bypass" capacitor around the emitter biasing resistor in one of the amplifier stages, causing degenerative feedback to reduce the stage's gain. This is a "Transistor Amplifier 101" level error, and was probably merely an oversight. I simply installed the requisite capacitor and immediately witnessed an absolute "quantum leap" in performance over our old system. I was the only person in the client's lab at the time; so I called Mr. Prater and said, "Don, you need to see this!" The device, called PERALS (Photon Electron Rejecting Alpha Liquid Scintillation Spectrometer), went on to win an R&D 100 (Then called IR 100) award; and the technology was ultimately transferred to Manfred Kopp, an I&C engineer who went into business for himself. I frequently use the following quote in my correspondence:

"May every young scientist remember... and not fail to keep his eyes open for the possibility that an irritating failure of his apparatus to give consistent results may once or twice in a lifetime conceal an important discovery."

In those days, it was often the I&C Instrument Technician who found and eliminated these "irritating failures."

In 1981, still with Millet's Analyzer Shop, I was promoted to "Engineering Technologist," a non-exempt salaried position that supplied tech support to the Instrument Technicians. (This was most fortunate for me because, shortly thereafter, the union went on strike. I had a pregnant wife and was in the process of building a house!) During this time I was asked to take a look at implementing a standard for personnel safety features on analytical and radiographic X-ray equipment at ORNL. The standard had been simply plagiarized from Y-12, and I quickly discovered that it was unworkable at an R&D Institution like ORNL. I finally "gave-up" and re-wrote the standard. I was immediately presumed to be the SME for X-ray machine personnel safety features and found myself "policing" these systems along with whatever else I was doing for many years afterward.

This job, as well as the Analyzer shop tech support job, began to require me to do a significant amount of circuit design and modification, so I was promoted to "Senior Engineering Technologist." I then reported to John Blanton, a General Foreman. This period of time gave me a great opportunity to work with researchers and put together all sorts of "special" little circuits to perform tasks that couldn't be done with off-the-shelf equipment. One of these gadgets even rated a patent disclosure. I was eventually promoted again to Principal Technologist, an exempt salaried position that was roughly equivalent to an entry-level engineer. Among the more interesting things that I did in this position was to help to devise a system of classifying and training our Instrument Technicians. The system that we put together was originally called BETT (Baseline Evaluation Testing and Training.) I also had a lot of fun actually teaching some technical courses as part of this process. This forced me to relearn an enormous quantity of material that I'd forgotten.

Around 1989, another opportunity was presented to me by then Department Manager, Don Miller. It seems that I&C Division's engineering sections were becoming far more interested in doing applied R&D than in actually supporting ORNL's other R&D Divisions. (Herein, in my humble opinion, lay a good many of the seeds of I&C's eventual demise.) They began referring to themselves as the "R&D Side of the House." Don perceived the need for a group of Field Engineers to take over the task of providing engineering support to the Instrument technicians and to ORNL R&D Divisions. This resulted in one of the best job situations that I've ever had. I was asked to become the group's leader. We named it the "Research Support Group" and populated it with ex-Instrument Technicians and Engineering technologists who had shown the potential for doing this sort of work. The group was, for the most part, immensely successful over the five-year span of its existence. It included Gerald Sullivan, AJ Beal, Mike Hurst, Bill Tye, James Bradford, Debbie Brophy, Vernon McClain, Scott Bruner, and Don Raby.

Following a two-year "stint" with a group in the "R&D side-of-the-house," during which time I worked on the new Badge Reader project and the MSRE Remediation Project, I returned to what was now called the Technical Support Section and assigned to supervise the Metrology Lab, the Fabrication Shop, the Steam Plant Instrument Shop, and a few other miscellaneous technical folks who had no other administrative "home."

This situation prevailed until I&C Division was dissolved in 2001; and I moved, with the Metrology Lab, to Quality Services Division."

It was during this era of significant changes in internal operations and external customer base that another unsuccessful attempt was made to strengthen the ATLC position by, to use union parlance, again addressing the relationship between Electrician and Instrument Technician craft workers.

Recognition of the rapidly changing political and scientific landscape was addressed in a lengthy I&C MMD strategic plan formulated in 1993. Several off -site meetings were held by Department staff to examine the challenges and develop a comprehensive response.

TSD SITUATION ANALYSIS – D.R. Miller

PRESENT SITUATION AND PLANNING ASSUMPTIONS
Present Situation for TSD - 1993

The internal groups and shops of the maintenance department (Technical Support Department) of the Instrumentation and Controls Division have been the primary source of support available to ORNL programs for electronic maintenance services. This has been the case since the earliest days of the Laboratory's existence. The Department was officially formed in 1978 by the (then) division director to address concerns of fragmented operations and personnel policy existing under the former structure. Since the shutdown of the HFIR and related increase in DOE focus on contractor operational processes, the Department has assumed a leadership position in the Laboratory compliance posture. More recently, it has become clear that absolute safety and the accompanying costs dictated by DOE are not the prudent path to take for the future. For the Department customers, technical merit and bottom line maintenance costs are rapidly becoming the key issue.

The I&C Division Position: The Division has just completed its first comprehensive research and development strategic plan. This plan is primarily focused on identification of key engineering strengths, and emerging technologies which complement those proven strengths and offer the opportunity for the Division to remain a strong and integral part of the laboratory's future. Division management has found the technical maintenance focus of the Department to be a desirable and necessary part of its future team.

Major Department Planning Assumptions:

- ORNL will remain a vital element in the overall national laboratory programmatic picture.

- Current ORNL programs, such as those managed by Analytical Chemistry, Chemical Technology, Energy, Environmental Sciences, Metals and Ceramics, Physics, Robotics and Process Systems, Research Reactors Divisions and the Security Department will remain essentially intact in some form.

- Many of the above programs will move to modernize existing instruments and systems before the year 2000.

- All new programmatic facilities will include significant systems or subsystems based upon new technology applications.

- Future Department management processes and systems will be required to meet a more "real world" validity check to be accepted. (later proved to be on target assumption)

– Computer work station installations will significantly increase each year through the year 2000.

– Pleasing customers and the one-stop shopping concept has been accepted as a primary business goal.

– By the year 2000 more than half of our supervisors will be retiring, requiring a significant increase in management focus on that process.

With the benefit of hindsight, the correctness of many of these assumptions becomes obvious. Notable exceptions include the change to an ORNL contractor that had few roots in the traditional business of this laboratory. Another was the unanticipated national political shift away from many traditional ORNL programs such as Energy and Environmental Science. Nonetheless, an indicator of the recognition outside ORNL of the success of these efforts is evident in this memo from Mike Cuddy who was on corporate president Clyde Hopkins' staff.

```
Date:      25-Apr-1993 11:28am EDT
From:      L Michael Cuddy
           CUDDYLM AT A1 AT OCB1
Dept;      7110
Tel Not    574-3332
```

TO: Use SH menu option to see recipients

Subject: Fwd: PERSONAL COMMENT REGARDING THE SELECTION TEAM VISIT TO I&C
File; READ   003370

Don, your I&C team is a major success story that we all are very fortunate to have had the opportunity to hear how continuous improvement in an empowered work team can work. As I listened to your folks describe the ownership of their work, I wanted to applaud. You and the I&C team made the tour a great success for all of us in Oak Ridge. I am proud to have worked with you over the last few years.

I know Clyde, Gordon, and Tom Young were impressed. Clyde told me he wanted to come back and hear more of what you are doing. Let's arrange a time for him to visit. If you will call Anne, we will be glad to arrange.

Great job. Please convey my sincere appreciation to the entire team. We need many more of our MMES team to hear what is happening in your world.

Addressees:

```
TO Miller, D R                    ( MILLERDR AT A1 AT OAX )
CC Trivelpiece, Alvin W           ( AVT AT STC06 AT UMCGATE )
CC Morgan, 0 Bill Jr              ( MORGANOBJR AT A1 AT OAX )
CC Stiegler, James 0              ( STIEGLERJO AT A1 AT OAX )
CC Calhoun, Anne Russell          ( CALHOUNAR AT DISTRIBUTION )
CC P01,1 AC:                      ( DISTRIBUTION LIST AT A1 AT OCB1 )
```

Memo from Mike Cuddy relative to Don Miller's Plans

During the early 1990s MMD was faced with a restructuring of the traditional in-house customer base and new demands from the Lockheed Martin Company to share expertise with other plants. Don Miller was asked to chair a new company committee with oversight of all five nuclear facilities then managed by Martin Marietta. He reported to Mike Cuddy of Y-12 and worked with representatives from ORNL's P&E Division, Y-12 Maintenance, Portsmouth and Paducah's Maintenance Divisions to study and share lessons learned about machine reliability. This task required Miller to both travel to other plants and to build the Machine Reliability Committee into a team able to develop policy necessary to unify and strengthen the five-plant maintenance capability and to deliver improved machine reliability.

The committee met face-to-face several times each year and, at a weekend retreat in a Kentucky State Park, worked on the overall plan. The work of this committee ultimately resulted in hosting a National DOE Equipment Reliability Conference in Knoxville. This two-day event included representatives from most DOE facilities around the country and many DOE staff. Also in this time frame, the I&C engineering sections were beginning to strongly market their services outside ORNL and became less interested in managing an internal service organization such as the Metrology Laboratory. TSD was restructured already and, with enthusiasm, took responsibility for the calibration work. At that point 90% of the metrology work was already being performed by TSD personnel. The new function required a top down review of metrology and creation of a formal document for day to day guidance. The preamble of that document read:

"1. INTRODUCTION

This document describes the methods and procedures for management of the Oak Ridge National Laboratory (ORNL) Instrumentation & Controls (I&C) Division Technical Support Department (TSD) Calibration Program. All personnel who have responsibility for calibration of equipment shall be familiar with the calibration program as described in this document.

0.   GENERAL: The primary function of the TSD Calibration Program is to ensure the measurement integrity of all instruments used to provide quantitative or qualitative data to meet program objectives and to ensure safe, reliable, cost-effective, and timely operation. The TSD Calibration Program includes measurement standards and equipment, technical personnel, TSD work centers, measurement equipment users, calibration data, and integrated planning combined in a structured program to ensure the reliability and accuracy of instruments, systems, subsystems, and equipment. The Calibration Program is a planned, systematic schedule of actions necessary to provide confidence that equipment used to

make measurements or quality judgments conforms to established technical requirements. It ensures measurement traceability to the National Institute of Standards and Technology (NIST) or other nationally recognized standards. Measurements in the mechanical, electrical, electronic, and nuclear fields are included. The contents of this document are reviewed and revised triennially by the calibration committee appointed by the TSD department head. The plan is distributed according to the current distribution list at the end of this document."

The Metrology Lab, while under the aegis of I&C Engineering Sections, was focused on seeking funding to do metrology R&D projects. This effort had always met with very limited success; the bulk of metrology R&D in the U.S. traditionally went to NIST (formerly called NBS), an organization far better equipped and positioned to do R&D in the metrology discipline. The distraction caused by attempts to do pure R&D had actually compromised the Metrology Lab's efficiency in serving the internal needs of ORNL's researchers. Under TSD, metrology staff were sent to short courses in metrology management and additional training was presented, in-house, to hourly staff and their supervisors, who were scattered around the Laboratory. The metrology management structure that resulted was highly successful and easily survived the transition to UT-Battelle site management.

The aforementioned Bob Effler was eventually chosen to manage the I&C Metrology Lab. Ironically, I&C Division was dissolved shortly after this. The momentum imparted by TSD's management vision to the metrology program, however, sustained the Metrology Lab into its subsequent incorporation into the ORNL Quality Division. In this organization it achieved internationally-recognized accreditation and became among the best metrology organizations in the entire DOE laboratory system. This notable achievement is a clear legacy of the sound management principles employed by I&C Division's Technical Support Department.

Events occurring in TSD following the retirement of Donald Miller were driven primarily by the choice for his replacement and external changes resulting from the choice of UT-Battelle as the prime contractor to manage ORNL. Dick Hess had little experience in the management of a large maintenance organization and no history in internal support and labor relations challenges. In addition Dan McDonald viewed the division as a permanent structure with assured budget and little need to continuously mend fences and build alliances outside the division.

Chapter 11

Contributions to Robotics and Remote Handling

William R. Hamel

The Beginnings

From its earliest days, ORNL was involved in wide ranges of experimental research associated with radioactive materials and often very high levels of radiation that required remote operations to protect workers. While the Laboratory has always been proficient in remote handling and operations, it was not considered a major remote techniques developer. This began to change in the 1970's as the lab took a leadership role in various aspects of nuclear fuel recycle and next generation nuclear power generation. In fact, during this period, ORNL became recognized as one of the leading laboratories in the world in applying emerging electronic and computer technologies to more effective remote handling systems. Technical staff members in the I&C Division led the majority of this research and development in concert with the Engineering and Fuel Recycle Divisions. Within the I&C Division organization, the I&C team was made up from individuals from both the Measurements and Control Engineering Section and the Research Instruments Section. The Robotics and Electromechanics Development Group was formed in 1985 and it grew into the Telerobotics Systems Section, in 1988, the first new section since the creation of the I&C division. The story of this era in which an array of talented and dedicated people literally pushed remote technology into new frontiers is best told by highlighting several of their important projects.

Advanced Fuel Recycle Drivers

In the 1970's the Laboratory was playing leadership roles in both the High Temperature Gas Reactor and Liquid Metals Fast Breeder Reactor fuel cycle R&D. Both of these fuel cycles involve significant amounts of fission products and other isotopes making remote operations in many phases a necessity. At this particular point in time, remote handling technology had been stagnate since the pioneering days of the Argonne National Laboratory's Remote Control Division. So, it was clear that remote operations and technology would play a major role, and that activities to evaluate the state of the art and to chart a course for the future would be needed. This responsibility fell on members of the I&C Division that were assigned to the programs. Over a period of about 15 years, the robotics and remote handling team led the develop of several new systems that were widely recognized for their fundamental contributions to a next generation of remote systems. This portion of I&C History is perhaps best told by describing these systems and relating some of the events that happened with them.

The CRL Model M2 Servo Manipulator

It was decided early in the breeder reprocessing program that the general approach would be based on full remote operations and that the first step regarding remote manipulators was to calibrate the state of the art in U.S. industry. The project to procure and develop a state of the art force-reflecting manipulator ended up being a joint effort between ORNL and the Central Research Laboratories. The Model M2 servo manipulator was based on an existing mechanical design but with modernized controls and electronics. ORNL was responsible for the majority of the controls and electronics from software through hardware. At this time, microprocessor technology had matured to the point that a distributed digital control architecture that would simplify wiring, improve control performance, and enhance operator control interfacing was more than feasible. Thanks to efforts of Joe Herndon, Lee Martin, Paul Satterlee, and Jim Phelps, the M2, shown in Figure 1, project was a roaring success. The system was able to give the operators perception of forces and loads on the order of 1 lbf while the arm could lift and maneuver payloads of 100 lbs. The team won a coveted IR 100 award in 1982 in recognition for the distributed digital control system. Over the course of a decade the M2 was used in many demonstrations showing how new remote manipulation technology could impact in-cell equipment design and maintenance. From time to time, the M2 was used to demonstrate robotic and remote handling concepts in other fields than nuclear. Figure 2 shows the M2 assembly a early structural assembly demonstrator for the space station program. NASA was interested in better understanding the relative performance between a robotics approach versus astronaut extravehicular activity. The truss assembly shown in Figure 2 had been assembled in space by two astronauts on the mission before the Challenger tragedy. Experienced operators using the M2 in the Remote Operations Maintenance Demonstration Facility in 7601 were able to successfully duplicate the assembly in about twice their time. This was considered impressively successful since the M2 tests were done in a 1 g environment with the equivalent of a single astronaut. These results had a strong influence on NASA to seriously pursue robotics as an integral part of the space station. Later, NASA funded ORNL to develop a prototype of a telerobotic manipulator that would point to the future directions of space robotics.

ORNL00232-84

Figure 1, The CRL Model M2 Servomanipulator System. Shown at the Operator's console are: L to R, H. Lee Martin, Paul Satterlee and Joe Herndon

The M2 was truly a remarkable system that was used to study many types of remote tasks. One summer sometime in the 80's, we were giving a demonstration to a Navy officer from their nuclear program. The officer noticed a wasp that had landed on the equipment that was in the field of view of the remote televisions. He asked Dan Kington, one of the best manipulator operators if he could swat the wasp with the robot arm. Dan immediately maneuvered the manipulator to the vicinity of the wasp and flipped it briskly into the next world with the end-effector. The officer was impressed.

The M2 did a fine job in showing the state of the art in manipulator design as well as showing the significant role that modern and digital electronics could play in making complex remote operations more effective. But, much work was needed in the area of mechanical design and that motivated a large effort to develop a next generation system that became know as the "ASM."

The Advanced Servo Manipulator

The Advanced Fuel Recycle Program was working on the next generation fuel reprocessing facility that would support liquid metal fast breeder reactors. There were many innovations in the designs being pursued and one of the most significant was that the plants would be as close to totally remotely operated as possible. This goal meant that advances would have to be made in a number of areas and particularly in the remote maintenance systems themselves. It was decided that a revolutionary approach to the remote manipulator systems would be taken as well. The advanced servo manipulator (ASM) was the cornerstone of this work. The ASM involved fundamental innovations in many areas including: 1) fully modularized mechanical slave manipulator design, 2) operator interface designed through human factors engineering, 3) state of the art digital control system that would support robotic automated operations, and 4) microwave -based wireless signal transmission. The ASM slave manipulators and the operator station are shown in Figure 3. The slave manipulators have colored anodized sections that are the replaceable

ORNL002140-88

Caption: *Figure 2 The Advanced Servo Manipulator slave system and operator control stations; Steve Zimmerman and Mark Noakes are at the control stations.*

modules that allow the system to be repaired in situ by other manipulators. The joint drive motors are located on the pod above the shoulder joint. Motor torque is transmitted from this pod through systems of coaxial torque tubes with bevel and straight gear interfaces between modules. The ASM contained over 200 precision stainless steel gears and was designed to minimize friction and maximize backdrivability. Dan Kuban lead the design team in the Engineering division and essentially all of the components were fabricated in the ORNL machine shops. This was some of the most complex and finest work ever done by the ORNL engineers and machinists.

The ASM control system was very advanced for its day. Position-position bilateral control was used to implement force feedback to the manual controllers which were kinematic replicas of the slave arms. Kinematic replica masters were used because there was some uncertainty that complete kinematic transformations could be computed in real time. The hardware was based on 16-bit processors in a multi-processing bus based architecture. The software was based on the FORTH operating system and language; over 50,000 lines of code were written. This design was pushing the state of the art in the early 80's with respect to complex embedded computing including extensive use of graphic displays. The control system was designed from the beginning to support teach/playback robotic execution of operator specified tasks.

The operator control station design also pushed the state of the art with extensive use of color graphic displays and voice inputs. The design of the consoles was based on established human factors principles and could accommodate from the 5th percentile female to the 95th percentile male. These facilities became a routine stop for ORNL visitors for many years.

About the time this system was going through debugging and beginning to be readied for performance trials, the fuel recycle program was throttled back into a shutdown mode because of the policy decision to not pursue nuclear fuel reprocessing due to proliferation concerns. One of the biggest career frustrations for the I&C team was premature termination of this work. Even though they had one of the finest robotics testbeds in the world at that time, they would not be able to explore its potential, nor have the opportunity to refine it.

Autonomous Systems

Around 1988 the DOE Office of Basic Science chose a team in the Engineering Physics and Mathematics Division to develop a basic research program in advanced systems engineering with a focus on intelligent systems. The Center for Engineering Systems Advanced Research (CESAR) was organized by Chuck Weisbin. Staff form the Telerobotic Systems Section in I&C became an integral part of the CESAR team and led the development of the HERMIES series of autonomous robots. Initially these mobile robots were small devices with rudimentary sensing and manipulation capabilities. The last was HERMIES III, which was much more sophisticated

ORNL8776-85   *ORNL04821-87*

Figures 3 & 4, HERMIES II (on the left with Bill Hamel and Chuck Weisbin) and HERMIES IIB (on the right with Chuck Weisbin)

with human sized manipulators, one of the first precision laser range scanners, and an on-board parallel computer system.

The HERMIES IIB robot was internationally recognized for its pioneering research on autonomous mobile task execution.

The robot manipulator on the HERMIES III robot was called CESARm. It was based on the ASM master controller manipulators but included an additional revolute joint at the based that made it one of the first kinematically redundant manipulators in the world. Scott Babcock led its development.

ORNL00550-90

Figure 5, The anthropomorphic scale HERMIES III autonomous robot

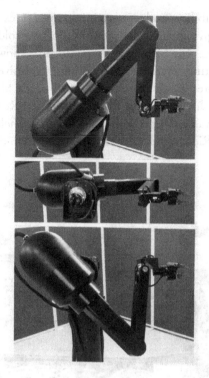

ORNL06767-86

Figure 6, The CESAR kinematically redundant research manipulator.

Shifts to Military and Space Research

In the late 1980's, the I&C Team sought R&D opportunities with other federal agencies to offset the reductions in the fuel recycle activities. Considerable work was done for the US Army and the NASA. Contributions are highlighted below.

The Soldier Robot Interface Project

In the 80's and 90's, the I&C Team worked closely with the US Army Human Engineering Laboratory (HEL) at the Aberdeen Proving Grounds. HEL was instrumental in introducing robotics into many Army missions. Projects ranged from using robots in field material handling and logistics to remotely driving unmanned ground vehicles. The Soldier Robot Interface Project (SRIP) was developed by the I&C team and a group at the Tooele Army Depot. It was a remotely operable all terrain vehicle platform that could perform feasibility studies pertaining to various types of missions/applications while giving particular attention to how soldiers would actually operate such systems. The SRIP mobile platform is shown in Figure 7. The SRIP was used to study material handling and explosive ordnance disposal tasks extensively. Later the Army allowed DOE to use SRIP in buried waste site surveying at both ORNL and Idaho sites. The SRIP operator interface was a compact stowable design consistent for Army field operations. Steve Killough and Brad Richardson were heavily involved in the SRIP.

ORNL00182-92

Figure 7, The Soldier Robot Interface Project mobile platform

The NASA Laboratory Telerobotic Manipulator

In the late 1980's Al Meintel from the NASA Langley Research Center was influential in shaping NASA's approaches to robotics. This was after the space shuttle remote manipulator that was developed by the Canadians. Al and others in NASA wanted the US to retain responsibility for the robotic systems that would be part of the Space Station. He was familiar with ORNL's work in robotics and remote handling and asked the team to support their efforts. The first substantitive project was a demonstration project in which the CRL

Model M2 system was used to assemble a prototype structural assembly so that the "robotic" results could be compared with an earlier demonstration that two astronauts had done in the space shuttle bay. Figure 8 shows the M2 at work in the 7601 high bay area. The positive results of this demonstration and the determined efforts of people like Al Meintel created a very positive and supportive environment for the creation of a new NASA robotics effort. The first step was the Laboratory Telerobotic Manipulator Project (LTM), which was intended to allow NASA to evaluate emerging technologies that position them to have the level of remote operations being used in nuclear applications in low earth orbit.

The I&C and Engineering teams that had worked together so closely and successfully embarked on the creation of new manipulator that would have a more dramatic form of modularization than the nuclear ASM. The LTM was designed as a 7 degree of freedom manipulator comprised of two types of 2 dof modules. The design involved many first time innovations. These modules were based on differential pitch yaw joints using roller traction drives rather than gears. The modules also included embedded electronics for signal conditioning, data communications, and joint/torque sensor interfaces. Fiber optic cables were used to transmit these signals from joint to joint. The electronics were smart enough to identify the module type simply by plugging it into its neighbor module. Figure 9 shows one of these modules and the embedded electronics. The gold rolling surfaces are the traction drives which were gold plated through a special deposition technique to evaluate this approach as a space vacuum lubrication approach.

The overall control system was derived from the ASM multi-processor design. It was built using the OS-10 real time operating system and the C language. It included many "hooks and handles" to support future robotics functions. The LTM was revolutionary in many respects and consequently was a very high risk project, which the team struggled to complete within budget. The final system is shown in Figure 9 just prior to shipment to NASA Langley. Unfortunately, the LTM never had the opportunity to seriously impact future NASA robotics in the manner once hoped for. Soon after it was shipped to Langley, the NASA once again subordinated the robotics aspects of the Space Station to Canada and thereafter not only was the LTM of little interest, the Flight Telerobotic Servicer Project at Martin Marietta was eventually canceled.

ORNL06457-86

*Figure 8, The CRL Model M2 Servomanipulator demonstrating
space robotic operations*

MODULAR CUSTOM ELECTRONICS ESSENTIAL TO
MULTISENSOR APPROACH WITH EMBEDDED WIRING

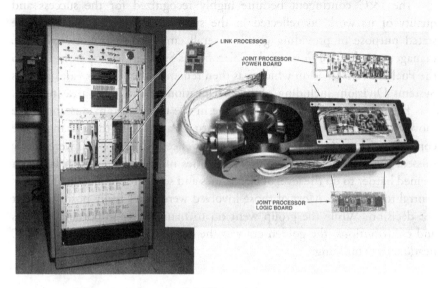

ORNL8015-89

Figure 9, The NASA LTM 2 DOF manipulator module.

ORNL0170-89

Figure 10, The NASA LTM System in Building 3500 during testing

The Big Move

The I&C contingent became highly recognized for the success and quality of its work, as reflected in the systems described earlier. For the stated purpose of providing greater overall emphasis on robotics, ORNL management decided to move the Telerobotics Systems Section from I&C to the Fuel Recycle Division which was then renamed the Robotics and Process Systems Division, including physical relocation to the 7600 area in 1990. It is fair to say that essentially everyone in the team was anxious about the move and its downsides. While folks eventually adapted to these changes, the comradery felt and the collaborations experienced while in 3500 were dearly missed. Even though the laboratory facilities in 7601 were much larger, it seemed harder to do the level of electronics and software engineering that was central to the work. Most of those involved were never really positive about the decision. While the group went on to many other important projects and contributions, the golden era was the days in Building 3500, the I&C headquarters building.

Chapter 12

Multi-Plant Secure Personnel Badge Development

Editor's Note: This was a project of the central management organization, Union Carbide Nuclear Division **(UCND), which at that time managed the operations of the three Oak Ridge plants plus the Paduca, KY plant, employing about 20,000 people. Thus the various writers in this chapter may have been associated with any of the four facilities.**

A Compilation of Several People's Recollections

From C.C Hopkins

C.C. Hopkins is a (retired) former CEO of the Union Carbide operations, responsible for all of the Oak Ridge and Paduca, KY Union Carbide Nuclear Division operations – Here is what he responded in an e-mail to Ray Adams:

"When Roger Hibbs became President of the [UCND] Company he expressed concern that the four facilities were too autonomous and had very little desire to work with each other, particularly in the administrative areas. Each facility had it's own Badge and Pass offices and had to contact each other for an approval of employees to visit, much like an outside visitor. When an employee transferred from one site to another, a physical examination was required at the receiving plant just like a new employee. In 1970 I transferred from Y-12 to K-25 and was asked to take a physical at K-25 by the medical organization there and I refused to do so. I had just had one at Y-12 shortly before the transfer. After a few days the problem was escalated up to Hibbs' office and he decided with me. This clearly identified some inefficiencies and ridiculous policies that needed to be examined.

Roger already had a strong desire to encourage us to operate as one Company instead of four. He wanted us to interchange people and "best practices" from site to site to strengthen performance of the Company. This change of policy and direction was the beginning of years of collaboration and support among the sites. He began having regular staff meetings with facility managers for the first time in the history of the company to help us learn about issues being dealt with at one site that might impact another and to provide help by one site to another to work crucial problems with the best talent in the Company.

Providing badges that looked identical and were identical to the extent possible was one of the first steps to support this new management philosophy and of course it brought some efficiencies. This is my best recollection about why the decision was made and Bob Affel was selected to coordinate the effort."

Ray Adams' recollection of the Badge Project

When Bob Affel was in the Central UCND plant security organization, he solicited proposals for a new Personnel Badge that would (among other features) be:

Really secure - non-copyable

Could be used by all plants (Paduca included)

Would notify a central security computer that access was sought at a portal, and would consult a central data-base for clearance.

Bear in mind that the three Oak Ridge Plants are physically separated from one another by about 10 miles, and that there are many (10 to 20) guard stations (portals) at each plant, oftentimes with different levels of security (and separate portals) within each plant.

The ORNL I&C Division responded to Affel's RFP, with a badge design based on the Wiegand Wire technology. In I&C, Cebert Mitchell proposed the overall badge design and Ed Madden's group was proposed to provide the badge readers.

The project was alloted to three primary groups:

1. The Badges per se – that needed to incorporate the security device, radiation monitors and visual ID. I&C employee Chet Morris handled the project from the standpoint of Badge design and fabrication.

2. The Badge Readers – that needed to communicate with the central computer and with the guard on station at a portal. I&C Group leader Ed Madden, handled the project from the standpoint of the badge readers, and testing of the badges.

3. The central computing would be handled within Affel's own security organization – Jim Simmons did the bulk of the programming.

The I&C proposal included the Badge per se; It was submitted by Cebert Mitchel and Ed Madden's group was proposed to provide the hardware for the Badge Readers. The proposal was submitted and beat a deadline that Affel had set. It included a prototype of the badges.

Chet Morris' Recollections:

I&C DESIGNS ADVANCED NEW SECURITY BADGES - (New Security Badges Use Wiegand-Wire Technology)

A. C. Morris, Jr., PE

In the early 1980's the US Department of Energy (DOE), Oak Ridge, needed a more secure personnel badging system for use at its four-plant operations. This requirement would eventually lead to developing and issuing over 20,000 new ID badges to all employees, with Robert Affel of the DOE Security Division heading up the overall project.

One overriding specification for the new badges was that an identification feature be included, unique in its coding for every individual, which could not be easily altered or duplicated. Several members from ORNL's I&C Division including Wilbur Allin, Tom Barclay, Dick Fox, Ken Knight, Cebert Mitchell, and Chet Morris formed a technical team for designing, developing, and producing this advanced badge system.

Additionally, DOE specified that this new personnel badge must: Display a recognizable color photo for each employee with his/her name in large letters underneath; indicate the wearer's security clearance level recognizable from a 50-foot distance; be sturdy and weatherproof to resist physical damage and outside exposure; and, be bonded into one integrated whole so attempts at alteration or reproduction would become immediately visible or detectable.

To meet the primary requirement of individual coding security, a number of available badge-encoding methods were reviewed and ranked by our I&C team. Included in these evaluations were printed bar-codes, hole-punched patterns, magnetic-tape stripes, dot-printed matrices, fingerprint-pattern readers, plus some other interesting (and sometimes nebulous) identification methods. One new encoding system that caught our attention early, and which was eventually incorporated into the new badge design, used a Wiegand-wire-coded insert that could be hidden and bonded inside each badge.

Basically, the Wiegand effect is produced in a small ferromagnetic wire (usually about 0.010" diameter by 0.50" long) that has been mechanically processed to produce a magnetically "hard" surface (shell) and a magnetically "soft" center (core). When such a wire is exposed to an alternating longitudinal magnetic field of an appropriate strength, the magnetic flux within this wire will "snap" quickly from one flux polarity to another. This abrupt magnetic flux jump is easily detected by a nearby sensing coil. Typically, a sense coil pulse of over 2 volts amplitude, lasting about 10 μsec, is obtained on each flux transition - a pulse that easily drives semiconductor circuits without amplification. Similar wires, ones made without the special Wiegand processing, do not produce these large snap pulses when exposed to such magnetic transitions.

For the identification coding practice, wires-with and wires-without this Weigand processing are arranged and glued, spaced side-by-side about 1/16" apart, along a 1/2-inch-wide plastic tape. Typically, from 25 to 50 such parallel wires are used per tape, and the particular pattern of "wires-with" and "wires-without" Wiegand-effect-processing will produce the individualized binary coding. When such wire-coded tapes are incorporated into a badge, and are run through a special Weigand-sensing reader, the number of possible and different binary code combinations becomes very large. For example, when one uses badges having 30-wire tape inserts, there are 2^{30} (or just over 1 billion) different binary codes available. And, because of the Wiegand-effect encoding, these codes cannot be sensed, read, or altered by any standard external read/write magnetic instruments.

Production of the new DOE badges proceeded using a final design having 7 separate layers, plus 2 inserts. A pressure/temperature manufacturing press cycle bonded all badge parts into an inseparable whole, six badges being made per press run. The internal Wiegand-wire tape was retained in a rectangular slot cut out of the badge's 0.020"-thick center Fiberglas layer. Another rectangular slot in a front vinyl layer served to position the employee's color photograph.

Initial manufacturing operations on this new badge had some interesting and often humorous moments. Too little press heat and pressure resulted in badges that immediately de-laminated and fell apart. Conversely, too much temperature and force resulted in hot liquid vinyl pouring out of our press, like an overfilled waffle iron. Through many trial runs we worked out correct heating and pressure combinations, eventually producing the required 20,000-plus badges on schedule. Wiegand-effect badge readers were installed at all plants and, after some field adjustments to equipment and on-line computer logging programs, we were soon able to pass verified employees through plant gate portals without undue delays.

There were some additional attributes related to this Weigand-wire badge design that we later came to appreciate - but only after the badges were issued and operating in field service. The Fiberglas center layer gave a very high mechanical strength to the badges, even when misused as ice scrapers on the frosty windshields of cars. Wiegand codes stayed intact over a -80 to +260 degree C temperature range, thus remaining unaffected when left in closed autos on hot summer days. Our badge strap design later allowed various radiation dosimeter elements to be attached behind the badge, without altering the basic badge functions. And most importantly, immunity of the Wiegand-coded badges to elevated magnetic fields was soon appreciated by many staff members, especially those physicists and engineers who worked in ORNL's high-magnetic-flux superconducting, cyclotron, and fusion-energy laboratories.

Badge Readers and Testing

Compiled from the 1980 I&C Division Progress Report - ORNL-5758

Ed Madden's group submitted the following reports pertaining to the Badge Reader project:

5.47.1 Evaluation of Badge Identification Components -- E. Madden, C. R. Mitchell, Y. H. Etheridge[14], G. W. Allin, R. G. Affel[15], A. C. Morris, and R. J. Fox

A program has been carried out to implement a new employee badge for the year 1980. The plan to acquire, evaluate, redesign, and develop suitable magnetic badge modules and badge identification hardware components for the new badge program required a long-range Instrumentation and Controls Division participation, as follows:

1. A DEC PDP-ll/04, disk-operated computer system was specified, purchased, and used to simulate portions of a site-wide employee badge identification controller to be used in the investigation of electronically readable badges.

2. Eight Z80 microprocessor-based, badge reader demonstration units were specified, purchased, and used in the badge module investigation. Each unit included a magnetic badge reader assembly, employee and guard alphanumeric display units, a video terminal display driver, a memory for downline loading of a data base from the host processor, a power supply and a key-locked housing.

3. Evaluation services and engineering recommendations were generated for suitable commercially supplied badge identification modules, suitable reading devices, and specially fabricated test badges.

4. An in-field badge and equipment evaluation test was implemented and installed at Portal 9 of the Oak Ridge Gaseous Diffusion Plant (ORGDP). The test involved approximately 250 employees, test badges, a local portal-housed computer, and two badge readers.

5. Three motor-driven assemblies were designed and fabricated for use in an in-plant acceptance testing program of commercially supplied magnetic read-head assemblies, magnetic badge identification modules, and the new employee badges. Two ten-pass, single read-head testers and one single-pass, ten read-head tester with appropriate electronic controllers were built.

6. A DEC PDP-11/34 computer and necessary peripheral devices were obtained from various Y-12 plant facilities and vendors, and the computer and devices were assembled and placed into use as a controller for an in-field test of both the new employee badges and the redesigned badge reader equipment.

The development magnetic badge modules were used at the Y-12 plant in the fabrication of over 20,000 new employee badges. The new badges were issued in April 1980 for use at all four plants of the Nuclear Division.

5.47.2 Badge Fabrication -- A. C. Morris, Jr., and G. W. Allin

Security badges were issued to all UCC-ND personnel at four plant sites in Oak Ridge and Paducah during the first quarter of 1980. These new badges were developed using a four-plant-matrix team of engineering, scientific, health-physics, security, and production specialists - each discipline contributing a definite and significant part toward the program's success. In pursuing a Department of Energy directive to issue and begin using the new badges by early calendar year 1980, the project personnel encountered a wide range of electronic, magnetic, adhesive, printing, and laminating problems.

The resulting badge design includes nine component layers combined with such inserts as an indium foil disk, a magnetically encoded identification module, and a low-fade color photograph. The nine badge layers (composed of vinyl, polyester, and fiberglass) are combined with the three inserts, and all components are adhesive bonded or laminated into a single unit by a laminating press that steam heats the components rapidly to 124°C while holding a uniform pressure of 1.034 x 106 Pa (150 psi).

Several distinctive features are incorporated into each new badge:

1. An unseen, 19.05-mm-diam, 0.381-mm-thick (0.750 x 0.015 in.), indium foil disk will alert health-physics personnel to any neutron or criticality exposure.

2. A magnetic identification module is encoded by a unique method that can be electronically read to admit or reject persons seeking entrance to restricted plant areas. Special badge readers and computer-terminal connections will be installed at appropriate gates or portals for this purpose. The magnetically encoded information is completely unaffected by use near cyclotrons or other sources of strong magnetic fields. -

3. Employee numbers exhibited on the rear panel of each badge are printed in an optical character reader format which permits reading with a suitable character-recognition wand. Numerous future uses (e.g., medical record files, library book check-outs, personnel files, parts check-outs, and insurance claims) are already envisioned for these machine-readable employee numbers.
    The new badge system involves only one badge per employee, replacing the previous system where two separate badges were alternately exchanged at the end of each quarterly monitoring period. Since all employees will now change their own dosimeter packets on this one badge, and no second badges need be stored, the amount of time and space needed by plant badge and pass offices for each badge exchange is significantly reduced.
    Over 20,000 of these new security badges were issued, including 1600 being used by persons who pass through portal 4 at the K-25 plant, where a prototype of a badge reader is being demonstrated.

5.47.3 Development_of Badge Identification Hardware -- E. Madden, C. R. Mitchell, Y. H. Etheridge[14], D. E. McMillan, and R. G. Affel[15].
    A program is being conducted to implement a computer-based badge identification and validation program for portal entry at the ORGPD. Both the Union Carbide Engineering Division and the Computer Sciences Division participated in the system development, acquisition, and installation planning. This program overlaps the 1980 Union Carbide employee badge development program described in the preceding two reports. The Instrumentation and Controls Division participated in this program as follows:

1. Six prototypic test badge reader and display units were designed and fabricated for demonstration and evaluation.

2. Prototypic employee badge identification hardware and software were modified and redesigned in preparation for badge equipment tests conducted at portals 4 and 6 at the ORGDP.

3. The electronic badge reader equipment was installed and used with a remote, computer-based controller in a badge identification evaluation program at ORGDP portals 4 and 6, with approximately 1600 employees participating. Newly issued, permanently assigned, Union Carbide employee badges were used in the test.

4. Six Intel 8748 microcomputer-based, badge reader simulators were designed, fabricated, and programmed for use in support of a communication loading test of the central badge validation computer located in building K1020 at the ORGDP.

5. A portal entry badge reader is being designed to implement the ORGDP final system requirements.

   This development program will be continued; installation of the central security equipment and all portal equipment for the ORGDP is scheduled for June 1981. Installation of similar equipment will be scheduled for all other plants in the Nuclear Division.

5.47.4 Security Module and Badge Tester -- G. W. Allin, E. Madden, and Y. H. Etheridge[14]

   A security module and badge tester was developed that automatically cycles encoded modules (either as-received or as-laminated into a security badge) across a signal-producing read head a preset number of passes under computer control.

   In the tester, the modules or badges are driven across the read head at a speed of 25.4 cm/s (10 in./s). This enables an operator to test approximately 1000 modules per 8 h shift.

   An optical character reader was also included as a part of the tester; it reads the employee number which is typed on the back of the badge in an optical character recognition (OCR) format.

   This tester device is also used to automatically enroll the employee number and its associated security number simultaneously into an appropriate computer memory for future security use.

5.47.5 Employee Badge Reader Simulator -- D. E. McMillan and E. Madden

An Intel-8748 microcomputer-based circuit was designed and fabricated to simulate an employee badge reader for deployment in a field test at the ORGDP. Front-panel selection allows simulation of either a valid employee badge entry, an invalid badge read, or an error read. The badge entry message is selected and transmitted to a security division host computer employee identification processor on a continuously cycled basis. After each transmission to the host computer, the simulator tests a reply message from the host for transmission errors and for proper message response. After a front-panel-selected delay period, the simulated badge read is retransmitted.

With the addition of a connector and a reprogrammed 8748 microcomputer integrated circuit chip, the simulator was converted to simulate a host computer for maintenance servicing of prototype infield service employee badge readers.

Use of a third, reprogrammed 8748 microcomputer integrated circuit chip permitted the simulator to serve as a portable field training aid when used with a prototype reader.

Since few hardware changes were required for conversion of the badge reader simulator to either a host computer simulator or a training aid controller, the maximum time required to redesign the simulator was 3 days.

5.47.6 Liquid-Crystal Alphanumeric Display -- E. Madden and C. R. Mitchell

A liquid-crystal alphanumeric display was built for the employee badge reader. The display accepts asynchronous EIA RS-232-C serial characters and displays up to sixteen alphanumeric characters. The input baud rate is programmable, and all standard baud rates from 110 to 9600 baud are available.

==

Jim Simmons' Recollections:

Note: there are probably too many details here for most people but this was my first job out of college and I'm still proud of the work that was done by everyone on the team.

I started working at Union Carbide Nuclear Division (UCND) in June of 1981, immediately after completing a BS in Computer Science. While officially hired by the Oak Ridge Gaseous Diffusion Plant (ORGDP, or K-25), I worked at Y-12 on the badge reader project from the start.

I worked in Bill McClain's group in the Computer Sciences Division, based at K-25. Bill's group specialized in systems development for mini-computer systems. When I started, Young Etheridge, another member of Bill's group, was already working on the badge reader project. While we were officially in Bill's group, we spent nearly all of our time working directly for Bob Affel, the Director of Security for UCND. We worked in a room near Bob's office in the Y-12 plant, that housed the development system (and central server) for the badge reader project.

By the time I started, the badges had been designed and everyone in UCND (about 20000 employees) had been given the new badges. The badge readers and controllers were nearly complete and prototypes of the software for both the readers and the central system had been written and demonstrated. About a year later we'd finished most of the software and had the badge reader system running at K-25.

The system was deployed at each of the UCND Oak Ridge plants that were geographically separated by as much as 10 miles. It ran from the date of its deployment, until the new operating contractor was in place on 4/1/84. It was eventually replaced with a newer computer system and much of the software was rewritten after I left the project in the late 1980's.

From ORNL's I&C division, we worked with Ed Madden, who oversaw the I&C work; Dave McMillan, who wrote the software for the badge reader controllers; and Cebert Mitchell, who did most of the design of the readers, controllers, and other specialized equipment we needed. There were others who also did a lot of work but I'm afraid I can't remember them all.

Work on the badge reader project was very much a team effort. Most of the initial overall design for the central software was done by Young and Bill, with much input from Bob, Ed, and Dave. While individual pieces of software or hardware may have been designed by a single individual, it was always with much input from the others on the team. In addition, a lot of input was sought and used from the individual Plant Security organizations, particularly the Visitor Control and Plant Shift Superintendent personnel who were the primary users of system.

There was also an understanding between all members of the team and our customers that it wasn't possible to design everything perfectly the first time. Any time it became obvious that a change to the design or implementation would significantly improve the system, there was always a willingness to make the changes and get it right.

This willingness to change actually helped to speed the implementation. It was quicker to design something, implement it, test it, then fix anything that really needed improving, if any, than to spend a long time on the design and implementation before knowing how well it would work. Bottlenecks sometimes showed up in unexpected places and we didn't spend a lot of time over-designing pieces that didn't need it.

That doesn't mean things were done hastily, though—there was always a lot of discussion during the design and many potential problems were eliminated before we started implementing anything.

The central server/development server was a DEC PDP-11/34, initially running the RSX-11M operating system and later upgraded to RSX-11M+. It had a fairly small amount of memory (256K, if I remember correctly) with two DEC RL02 (10 MB) disks. An RL 02 disk was a removable disk with 1 platter about the size of a large pizza. All work on the systems was done through DEC VT-100 type terminals connected through serial ports. In addition we used Emulex controllers that could each handle up to 64 additional serial ports to talk to the individual badge reader controllers.

The servers put in place at the three Oak Ridge plants were DEC PDP-11/44s running RSX-11M+, each with 2 Emulex controllers (128) ports and 2 RM03 disks. The RM 03s were removable disks about the size of a large round cake, with multiple disk platters that held 67 MB each.

The servers communicated with the badge reader controllers, at each of the (up to 25) employee portals, each manned by a security guard, through serial connections.

A badge reader controller was a box about the size of a large book. It had a serial port for communication with the server, plus ports for up to 2 badge readers and one for a printer, and connections to control door locks and monitor various switches. Each badge reader port could talk to both a keypad and a badge reader, and each badge reader contained a 20 character alpha-numeric display (either LED or LCD).

There were two types of keypads. One was a standalone one that looked like a phone keypad but with computer keyboard size keys, normally used by guards to assign temporary or visitor badges. The other was designed to be part of a badge reader. It had a row of keys that fit directly below the display on the reader. It was used for entering an individual's PIN (Personal Identification Number) after reading a badge at certain high security locations. This keypad didn't have labels on the keys. Instead the numbers 0 - 9 were displayed above the keys, either in order or in a random order. For locations where the keys were scrambled, we had special displays that were designed so you couldn't read them unless you were looking directly at them. We had this scrambled PIN entry working for years before I ever saw anyone else advertise anything similar.

To the badge reader computers (servers), a badge was just the pseudo-random number represented by the Weigand module it contained. Each module contained a series of Weigand wires that had been punched with a hole (similar to that made by a punched card reader). The hole cut either the top or bottom of the wire. When the badge was moved through the reader, it went past a strong magnet, then past a read head. As each cut wire went past the magnet, it generated either a positive or negative pulse. The reader interpreted these as a series of ones and zeros - a binary number.

The binary number actually contained some extra bits- a start and stop bit (one was a one, the other a zero) so you could tell which way the badge was read, plus some bits that made up a checksum. The badge reader/controller was smart enough to count the bits, check the checksum, reverse the number if the badge was read "backwards", then remove the extra bits to come up with what we called the badge id. The badge id was small enough to fit in a single 32-bit integer, which made it easy to handle in the database.

Note: when I talk about the "database", we didn't use a formal database system on any of the machines. DEC did sell one for the machines, but in our limited testing it was so slow it was basically unusable. We used DEC's indexed file system (Record Management System, or RMS), once RMS had improved enough to be fast enough for us.

A very important design consideration was that the badge reader servers handled all badge verification. On the back of the badge the user's badge number and clearance were present, but they weren't used to do access authorization.

When a badge was read, the badge ID was sent to the server. It looked up the badge in the database, identified the user it was assigned to, verified that the user was authorized to enter through a particular portal (a particular reader), and sent a message back to the badge reader displays and log programs. If the user was authorized, the message was just the user's name, a company abbreviation (for non-employees and visitors), and clearance and it went to both the user's and guard's display, along with a single beep. If anything was wrong (badge misread, user not authorized or any other problem), a "SEE GUARD" message went to the user and a message describing the problem went to the guard's display, and the reader beeped multiple times.

As an additional check for forged badges, the guard is supposed to check the name and clearance on the display and compare it with that on the badge. In some cases the guard just tells the person to re-read the badge. This happens most often if the user ran the badge too slowly or it wasn't flat in the reader.

One of the messages the guard could receive is "ESCORT REQUIRED". If the user is authorized into the area but requires an escort, the guard verifies the escort is there and then has him read the badge again. The server recognizes the two reads of the same badge in a row and sends the user's name and clearance back the second time.

The systems we were using were mini-computers, and as such we had to modularize things as much as possible. No individual program could be too large or it wouldn't even run, and with shared code it was far more efficient to run multiple copies of the same program, each handling a smaller portion of the work, than to run one large program that did everything.

The server software was written in DEC's FORTRAN, primarily because it was the best supported language on the OS. C was relatively new and not supported by DEC at that time. Though considered by most to be a scientific programming language, DEC's version had a lot of support for making low level Operating System calls and supported indexed file access (RMS).

It wasn't a well structured language though, and we eventually used a free pre-processor called RatFiv instead. RatFiv was an improved version of RatFor, or Rational Fortran. It gave us structured language elements like "if-then-else" statements, primarily, and made the code much more readable. To compile a routine written in RatFiv, the source was run through the RatFiv pre-processor first. It generated standard FORTRAN code which was then compiled.

We were limited by the linker in the length of names we could use for programs, subroutines and global variables.

For example, each database file had a two letter abbreviation. We had a "Badge Information" file that contained data on the badge modules we had requested or used. It was the "BI" file. There was a READBI routine that you could use to read a BI record into a global BI record (defined in an include file). It took a few parameters. One specified the type of read - either sequential (read the next record), first (read the first record in the file), or a letter indicating which key (index) was to be used to look for the record. For indexed reads you also specify the key to look for (for BI it was just the "badge ID"). Finally there was a status parameter that told you whether it found the record or hit end of file.

Many routines were required for such things as updating the "BI" file, searching it for visitor control purposes, and for other file maintenance purposes.

The main workhorse program on the servers was the badge reader task. When started, it took its task name and looked it up in a configuration file to see what badge readers it should monitor. Each copy normally handled about 5 readers but was written to handle up to 10. With testing we determined 5 was about the optimum number, considering that it could only process a single message from one reader at a time. By spreading the busiest readers across various copies of the program, we could keep the overall response fairly level.

The configuration file identified the particular serial port each reader controller was connected to, along with how many badge readers were connected, whether there were keypads or a printer, etc. It also identified the portal number and any special access requirements for the portal (clearance required, security level, etc.).

When the server was first started, a central task went through a configuration file and started all the badge reader tasks. The badge reader tasks read their configuration files and start talking to the badge readers. They also start several other processes, as needed.

One of these processes handled logging for the entire system. Anytime a badge was used or a message generated that needed to be logged, the badge reader task (or other program) would send it to the log program. The logger simply added a date/time stamp and wrote it out on the log file.

Unlike other files in the system, the log file was a simple sequential file. In early tests we quickly determined that RMS couldn't update indexes fast enough to keep from bogging down the entire system. In order to be able to quickly search the log files after the fact, we wrote programs that ran periodically to simply create sorted versions of the log files that could be searched using a binary search.

We also designed the system to handle visitors and lost or forgotten badges. For visitors, Visitor Control entered information on the visitor in advance, including name, company, clearance, visit dates, and which portals the visitor could enter. The system then added an entry to a file indicating at which portal the visitor's pass needed to be printed, then started a program which handled printing the passes.

This program read the database, and built up the passes in memory for the various portals. It then sent messages to the appropriate badge reader task to send pieces of the form to the printer attached to the reader controller at the portal. Once the form was successfully printed, it removed it from the queue.

There were limitations with the badge controllers (primarily due to the limits of the hardware at the time) that allowed them to handle only a single message (sending or receiving) at a time. Any time they were sending or receiving, they couldn't read a badge. So as a message started, the display on the reader was changed to PLEASE WAIT. Once the message was sent or received it changed back to the date and time. If someone tried to read a badge at this point, the controller would normally get just part of the badge id. In which case it notified the server to send a "PLEASE TRY AGAIN" message.

To make sure the controllers were working, if the server didn't receive a message from the controller in 20 seconds it would "time out" and send the date & time message to the display. This kept the display reasonably up-to-date and kept checking communications. In addition, the reader would automatically display a "SYSTEM DOWN" message if it didn't get a message every minute or so.

When visitor passes are printed a piece of the pass was sent with the 20 second time out message. This meant it could take several minutes to print an entire pass but it also meant it wouldn't even try to send anything if the reader is busy (people coming in less than 20 seconds apart). This actually worked very well since the visitor passes were usually entered into the system hours before the visit.

When a visitor arrived, the guard would look for his/her printed pass and confirm the identify using information printed on the pass and the visitor's drivers license or other ID. Then the guard picked up a random visitor badge (of the appropriate clearance level) and ran it through the guards reader. The badge reader then started a series of questions about the visitor that the guard used the keypad to answer. When everything is confirmed (including the visitor #, visitor's name and clearance), the badge was assigned in the computer and handed to the visitor. The visitor then runs it through the entrance reader just like anyone else.

A similar system was used to handle people who forget their badge. When someone identifies themselves and says they forgot or lost their badge, the guard gets the appropriate type of temporary badge, runs it through his reader, and answers questions that include the user's badge # and a verification of the user's name. Once complete, the badge is assigned to the user and it is good for so many hours, usable just like his normal badge.

At the same time the temporary badge is assigned, however, the user's normal badge was marked as "forgotten" in the database. When the user returns the next day with his regular badge and runs it through the reader, he gets a "SEE GUARD" message. The guard gets a message saying "FORGOTTEN BADGE". The guard then takes the badge, verifies the picture, and runs it through his reader. This resets the badge and also un-assigns the temporary badge if it is still assigned.

Overall, I believe the system worked very well considering the limitations of the hardware of the time. Of course, I'm sure many things would be done much differently today, but the basic design principles still would apply.

Chapter 13

Instrumentation and Controls Engineering – Process Instruments 1954 and Onward

Ray Adams' recall

I began my employment in the Instrument Department January 2nd, 1954. Having been formed only a year earlier, the I&C Division, for which no Division Director had yet to be named, consisted of two departments with several Sections. In the newly formed Controls Department, headed by E. P. Epler, I do not recall any Sections, but as the Instrument Department was formed from an earlier existing department under the Plant & Equipment Division, there were several Sections. Two large Sections were the Electronics Section and the Process Instruments Section.

Early years of the Process Instruments Section 1954 and Onward

I was hired by Ed Shipley and Charlie Harrill, to report on January 2, 1954, having received a transfer from the K-25 Instrument Engineering Department. I had started work at K-25 on July 31, 1951 and had become acquainted with many of the ORNL I&C Engineers at meetings of the Oak Ridge Section of the Instrument Society of America (ISA).

Process Instruments – Charles A. Mossman's Section

Early (1954) Members of Mossman's Section
As to the history of ORNL and The I&C Division, Alvin Weinberg formed the I&C Division in February of 1953. It included the Instrument Department under Charlie Harrill and a newly formed department of Reactor Controls, under E.P. Epler (who had previously worked for Weinberg's Physics organization. Cas Borkowski was persuaded in February of 1954 to

become the Director of the newly formed I&C Division, and he brought his instrument group from the Chemistry Division into the I&C Division. Both the Reactor controls Department, which had offices on the 2nd floor of Bldg. 4500 (in the attic back of the Executive Conference room) plus the analog computer (in E corridor, on the first floor of Bldg. 4500) and Borkowski's group (B or C corridor) of Bldg. 4500 remained in Bldg. 4500, until the new addition to building 3500 was completed, in 1960. E.D. Shipley (who brought two lunch boxes filled with strawberries and cream the Spring and Summer of 1954, and who ate in the lunch room of 3500) was acting Director of I&C. He retained his title of Associate Laboratory Director.. A bit later, he got interested in Fusion Energy and I think he moved to the Fusion Energy Division in Y-12 ,to work in that area when Borkowski agreed to be the I&C Division head.

The Instrument Department at that time was organized into a Process Control Section (the one I was in), an Electronics Instrument Development Section (Run by Frank Manning), a Radiation Detection Section (run by Roland Abele), a small mechanical development section (run by Bob Tallackson), a drafting room with about five draftsmen (run by Mark Bowell), and a Maintenance Section (run by Bill Ladniak) - all of the Instrument Technicians were based in the Maintenance Section. There was a High-Voltage section to support the Van deGraaf & Cochroft-Walton accelerators. That I&C group was run by Gene Banta, who was succeeded by Jim Johnson The total of the members of the I&C Division in1959, was about 200. In the Instrumentation and Controls Division, the parent organization of the Instrument Department, there were Basic Instrumentation, Sherwood Diagnostics, Instrumentation for Chemical Research and the Controls Department. These groups totaled about 35 persons, giving the I&C Division a total of about 235 people.

Shortly after I joined the group in 1954, my boss Warren Brand left ORNL (His company – Fischer Porter strongly supported the Instrument Society of America [ISA]) , and he shortly became National President of the ISA). At ORNL, a new I&C Group Leader had to be chosen. Charlie Mossman had become the de-facto group leader, but the Department Head Charlie Harrill, chose Bob Tallackson (the former leader of the mechanical design group) so Mossman and Tallackson in essence became co-group leaders. Tallackson soon moved on and I lost track of him - Mossman had always been (for me) the real group leader, after Warren Brand left.

I remember in that group, there were Bernie Lieberman, Tom Gayle, P.P. Williams, Steve Lisser (whose desk Mossman assigned me to use, as Lisser worked the night shift during the start-up of the OREX Pilot plant) and he only needed his desk at night so I could use it in the daytime. Lisser was not overjoyed at this [temporary] event, but bore it with his usual aplomb and we later became good friends. Other engineers in the group were Bill Greter, Larry Chase, John Horton, Steve Hluchan, Joe Lundholm, George Ritscher , who ran the Standards Lab, and Frank Potts.

I worked on the start-up of the OREX pilot plant - to separate lithium isotopes. Its control panel was of a new design that had a graphic outline of the plant in which (for example) the level and other control recorder/indicators were physically located in a depiction of the process vessel, and the control pumps and valves (or the switches to operate them) were located in lines on the panelboard that indicated the actual plant piping. The instrument panel really looked like a process flow sheet.

ORNL 11979

C.A. Mossman standing in front of the control panel of the OREX Pilot Plant – a new type of panelboard, a Graphic Panel.

The plant (the process flow sheet) a product of the ORNL Chemical Technology Division) employed an organic solvent and mercury amalgam of lithium in a counter-current flow through pulsed columns. Notwithstanding the snazzy look of the instrument panel, the OREX pilot plant process lost out to a Y-12 plant competing design (COLEX) that I think, used mixer-settlers (instead of pulse columns) as the counter-current flow contactor. Nevertheless, the innovative design of the instrument panels and rack construction was used thereafter, throughout ORNL for years to come, with or without the graphic design concept.

The engineers in this section worked on individual projects, when they weren't working on a large project like the OREX plant, that engaged Bernie Lieberman and Steve Lisser to do the instrument application engineering.

I recall that John Horton was an expert in the application of the L&N Speedomax recorder, that Joe Lundholm knew a lot about applications of the Brown Recorder (It was so-called because it had been a product of the Brown Instruments Division of Minneapolis Honeywell, but it was painted black). Projects at this time were mostly associated with one individual. Frank Potts did a Temperature Control system for Chemical-kinetics research (for Sheldon Datz in the Chemistry Division), as well as working on a High-Temperature adiabatic Calorimeter for testing fused salts. Another of Frank Potts' projects involved mechanical vibration studies. John Horton worked on Metallurgy Div. Creep-Test instrumentation, and Tom Gayle did such things as a flow recorder/sampler for 4500 Area Liquid Waste, as well as testing the precision of Low-Pressure Gas Regulators. George Ritscher, in addition to being the "Standards Lab." engineer, worked on a project associated with measuring potentials and currents of arc-welding apparatus for the Metallurgy Division, as well as instrumenting a rocking-bomb experiment for the Chemistry Division that was installed in the MTR reactor in Idaho.

The Standards Lab had a stock of high quality measuring devices that could be checked out by qualified individuals to check that measurements (mostly dc and ac voltage, current and power) were made to high standards. This Lab was a staple of the I&C Division and in later years had its own research agenda that included high precision temperature standardizations.

In addition to the needs of the ORNL research divisions (Chemistry, Physics, Metals and Ceramics, etc.) for instrumentation and controls expertise, there were in those years several large projects that required an integrated approach to measurements and controls. The Aircraft Nuclear Propulsion (ANP) project, the Homogeneous Reactor project, and the Tower Shielding Reactor (TSR) project were some of these. The ANP project accumulated a group of engineers and physicists that were solely dedicated to that effort. The Homogeneous Reactor project operating out of the Reactor Experimental Engineering Division (REED) organization had a captive group of I&C engineers, some of which were members of the REED Organization, and some of which were members of the I&C Division, but they were physically located at the Y-12 plant under the leadership of Don Toomb. The TSR project (located in a valley to the South of ORNL proper) drew instrument engineering expertise mostly from the main I&C organization.

Some examples of the projects that I&C Members worked on are shown in the next few illustrations.

ORNL41669
Powder Metallurgy Lab Furnaces and Controls

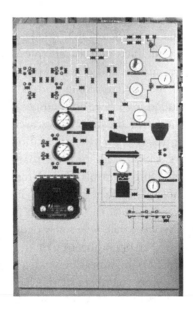

ORNL17493
Control Panel for the Fission Products Pilot Plant

The I&C Progress Reports for the years 1956-57 carry extensive reports of the Homogeneous Reactor Test (HRT) Instrumentation, written by Don Toomb, who by that time had transferred from the REED organization, to I&C. Don reported on the work of the several engineers in the HRT group, physically located in the Y-12 plant in the Reactor Experimental Engineering Division (REED).. Those engineers were: (1957) Syd Ball, Earl Bell, Arville Billings, J.R.Brown, Dell Davis, Glenn Greene, J.D. Grimes (on loan from TVA), Joe Gundlach, P.G.Herndon, Bob Moore, R. M. Pierce (also on loan from TVA), Jack Russell, and Harry Wills. The control panel for the Homogeneous Reactor was designed by that group in the I&C Division at Y-12 and is shown in the picture below.

ORNL17490
The control Panel for the ORNL HRT

By mid 1957, Steve Lisser and I completed the HRT Chem-Plant instrumentation application and control panel. The HRT Chem-Plant was designed to extract, on a continuous basis, the fission and corrosion products formed by the operation of the HRT. A paper written about this project, won a "best paper" award from the Instrument Society of America[16]. A picture of the control Panel for that process is shown here.

ORNL40054
HRT Chemical Processing Plant Control Panel

The instrument panelboards of those days all had the modular construction that the mechanical group of the I&C Division had helped design. Some of those instrument panels were a "graphical representation of the process under control," and some were not.

Later, (by 1966) what eventually came to be known as the Process Control and Instrumentation (PC&I) Section evolved a bit, and grew to include several groups. There was a Development Group, a Systems Design Group, a Data Processing & Analysis Group, and a Fabrication & Maintenance Group.

The I&C Progress reports from 1966 onward reveal work on many projects, such as these 1966 entries:

Data Processing & Analysis

    DEXTIR System expanded; Hyland & Adams

    MSRE Computer; Burger, Martin (incl programming)

    T/C Table smoothing; Simpson and Adams

    Work Order Labor Summary; Simpson

Process Insts & Control Systems

Thorium Utilization; Lisser

Nuclear Safety Pilot Plant; C.Brashear & B.C.Thompson

Transuranium Proc. Plant; Cochran

Project Salt Vault; Herskovitz & C.C.Hall

Fixation of Radioactive Waste in Asphalt; Toucey & Sliski

Tantalum Process Valve Development; G.W.Allin, Bernard Lieberman & H.J.Stripling, Jr.

Containment Research; Earl Bell

PC&I Planning; Lisser, Reyling & Simpson

Resistance-Type Liquid-Metal Level Element Development; N.H.Briggs, G.W.Greene & R.F.Hyland

Process Instrumentation Development

Power proportioning Device; W.R.Miller

Basic Control Unit; W.R.Miller & L.H.Thacker

Furnace Program Generators; W.R.Miller & L.H.Thacker

Furnace Burnout Monitor; J.T.Hutton

Multipoint Recorder Programmable Range Converter; J.T.Hutton

Graphite Evaluation Program; L.H.Thacker & W.F.Johnson

Nuclear Safety Pilot Plant Fog Detector; L.H.Thacker

Pnuematic Scanning System; W.R.Miller

Microsphere Production; W.R.Miller

Aids For Instrument Development; W.R.Miller & L.H.Thacker

Automatic Differential Thermal Analyzer; L.H.Thacker &   W.R.Miller

Electrolytic Polishing Controller For Electron-Microscope   Specimens; T.M.Gayle & J.T.Hutton

Emergence of Desk-Top Computers

In about 1983 a number of changes in the PC&I Section began to take place. C.D. Martin had for several years moved on from taking Ray Adams place as the head of the Digital Processing and Systems Analysis group, leaving it to be run by Jim Jansen as the Real-Time Computer Group, C.D. Martin was by then head of the Research Instruments Section, formerly headed by Frank Manning. About then, Ray Adams founded the Microprocessors and Desk-Top Computers Group in Charlie Mossman's Section (PC&I) in response to the sudden influx of personal desk-top computers throughout the entire laboratory.

That Microprocessors and Desk-Top Computers Group provided support for ORNL researchers needing hardware and software support for the thousands of desk-top computers which ORNL researchers and clerical staff had begun to use. That group almost became the support organization for the whole of the Union Carbide Nuclear Division's (UCND) desk-top computer effort. There were even a number of high-level meetings with UCND VPs and a desk-top computer support organization was briefly set up, prior to that support structure being taken over by the UCND Central Computing organization.

Then in April of 1984, Martin Marietta Energy Systems corporation became the ORNL operating contractor, and in September of 1986, Adams took early retirement to teach in the Electrical Engineering Department of the University of Tennessee at Knoxville.

Real-Time Computation – The Real-Time Computer group expanded its operation and (still today, in the ORNL MSSE Division) is a unique organization devoted to the acquisition and processing of large amounts of data, by computer. Typical of the sorts of systems they create is the following description of a project done for the U.S. Navy.

The Sound Of Silence

Editor's Note: This article appeared in the ORNL Review, ORNL's prestigious journal of technical papers. Vol 128 No 4 - Summer '95, It is reprinted here in its entirety, with only minor changes. All of the photos appeared in the original article, but their ID's have been lost. The author, Bill Cabage, is a lead author/editor in the ORNL Office of Communications and External Relations.

Bill Cabage

Silence is the hallmark of the United States Navy's nuclear submarines. Because they stay well out of sight below the surface of the water, enemies can detect them only by the sound they make. Consequently, they are designed to ply the seas as quietly as possible the aquatic version of stealth technology. Many of those quiet technologies are closely guarded secrets.

This new technology can detect sounds below the ocean's background noise.

The Navy prefers to call its subs "quiet underwater weapons platforms." What little noise they do make must still be monitored in operational tests, both to improve designs and to ensure that the craft are operating to specifications. The testing requirements presented engineers with a challenge: how do you measure the sound of something if it makes almost no sound?

The U.S. Navy has relied on ORNL's Instrumentation and Controls Division's expertise in electronics, real-time computer applications, and systems integration to develop state-of-the-art acoustic measurement systems. ORNL has been working with the Navy on a special program for the past 6 years called AMFIP II the second phase of the Acoustic Measurement Facilities Improvement Program. According to Randall Wetherington, who headed the program in the I&C Division, ORNL has devised ways to "measure the undetectable" for its sponsor, the Naval Surface Warfare Center's Carderock Division.

The system uses several arrays of hydrophone sensors--laceworks of underwater microphones suspended from buoys--to gather sound from the craft as it goes by. Computers process the sound signatures to extract the signal from background noise. "Waves and wind cause an ambient noise level in the ocean," Wetherington explained. "This new technology can detect noises below this ambient level. It's like a TV satellite dish in that the array, like the big dish, focuses the energy, which amplifies the sensor signals. As you add sensors, with their placement based on sophisticated math and geometry, you get more signal gain over a large bandwidth."

The USNS Hayes is a specialized ship that is home to the instrumentation systems used in the Navy's acoustics tests.

Much of the signal processing hardware and software for I&C Division's AMFIP II system is currently aboard the Navy's laboratory research ship the USNS Hayes, which is currently ported at Cape Canaveral, Florida. When test runs for submarines are scheduled, the Hayes sails to the test course set up in Exuma Sound, which lies in the middle of the Bahama Island chain. There the arrays, which have more than 1000 individual hydrophone sensors each, are placed to pick up sounds from the vessels as they pass through the course. Long umbilical cables connect the arrays to the rest of the system, which is located aboard the Hayes.

The work with I&C engineers often had to be done at night. The Navy's test course is located in the Bahamas within a corner of the fabled and mysterious Bermuda Triangle. Researchers in the AMFIP II program like to point to this fact, although few expressed reservations about working after dark.

The radiated noise emanating from the Navy's submarines, called the acoustic signature, is normally low enough to be masked by the natural background noise of the ocean in a conventional hydrophone setup. The next generation of nuclear submarines the SSN 21 Seawolf operates below this natural background, requiring the world's most advanced underwater acoustic measuring system to ensure that its acoustic emissions remain below design limits and thus stay hidden beneath the vast ocean.

Like a number of ORNL researchers involved in the AMFIP II project, Randall Wetherington (left) and Andy Andrews have made a number of trips to the Bahamas' Eleuthera Island, which lies near the Navy's test course in Exuma Sound.

Staying hidden is part of the submarine's job. Acoustics is one of the few ways these ships can be found.

The initial problem, says Andy Andrews, deputy program manager in the I&C Division, is how to measure something that makes virtually no sound. "Staying hidden is part of the submarine's job," Andrews says." Acoustics is one of the few ways these ships can be found. The craft in the Trident class can be the length of two football fields and not make any noise as they go by. Our systems help characterize the normal acoustic signature of the vessel as it runs through a course of instrumentation. Our system is not classified, although the data it generates are. If you know what a vessel's acoustic signature is, you could develop ways to detect it."

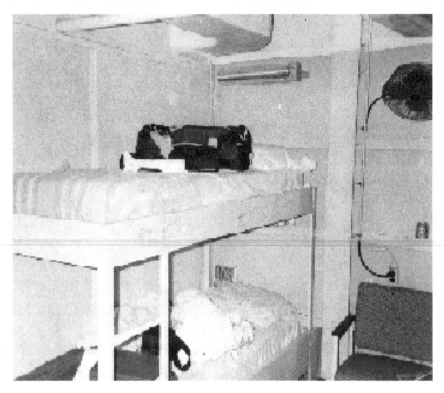

As are many a sailor's, shipboard accommodations aboard the Hayes are spartan.

Wetherington likens the AMFIP II system's acoustic feats to a familiar scenario: "It's the Tennessee-Alabama game at Knoxville in the last minute of play. The score is 28-27, and Alabama, who has just scored to come within one point, is lined up to try for a go-ahead two-point conversion. Imagine the crowd noise from about 100,000 people. Now, sitting near the 50-yard line is an English professor with a heart condition who, in times of stress, calms himself by reciting aloud Edgar Allen Poe's Annabelle Lee. "This acoustic technology could pick the professor's poem out of all of that crowd noise."

Electronics and Hot Rod Chips

The ORNL researchers' efforts have focused on three main areas--telemetry and underwater electronics; beamforming, or signal processing; and system integration, or, more simply, making it work. More generally, they've designed and built electronic instrumentation and developed a processing system that takes the data and generates information in a usable form.

AMFIP II's instrument arrays feature numerous hydrophones and

204

sophisticated underwater electronics. The telemetry system was a challenge to the I&C AMFIP II team because of the large number of signals from the sensor arrays that must be acquired, conditioned, and transmitted.

These signals have a very wide bandwidth and must be transmitted over a long distance. The system required the identification and use of emerging technologies that arrived on the market at the same time that the design effort was initiated, making AMFIP II truly state of the art. The AMFIP II team had to clear technical hurdles presented by power distribution, heat dissipation, high-speed and high-resolution signal digitization, and very high-speed data multiplexing combining information cascading in on multiple lines into a single line suitable for fiber-optic transmission.

Schematic Representation of the AMFIP II sensor array,

Resembling a huge, mechanical Portuguese man-o-war, the AMFIP II sensor array is hoisted over the water by technicians aboard the Hayes. The latticework of hydrophones more than 1000 for the system is strung above pressure vessels containing instrumentation.

Our success in identifying new technologies produced significant cost savings for the Navy.

"We used a 'hot rod' multiplexer chip developed by DARPA that combines information on 40 lines into a single line," Andrews said. "Our success in identifying new technologies produced significant cost savings for the Navy. For instance, we convert signals from the hydrophones to digital data almost immediately, which gives us very high-resolution signals, and then transmit the data by fiber-optic cable to the test ship. We identified a $20 digitization chip that will do tasks that used to cost $3000 per channel.

We built three systems with more than 1000 channels each one channel for each sensor."

Along with combing the marketplace for available electronics, the AMFIP II team also oversaw a large subcontract effort required by the construction of around 1300 multilayer printed circuit assemblies of 50 different designs, 22 equipment chassis, and countless cables and test fixtures. All of this, as well as automated test equipment to verify that they were all working properly as they were being made, had to be designed, fabricated, and integrated into the final system.

A major advantage of the system is that sensors can be monitored and diagnostics can be performed aboard ship.

When something breaks, as is apt to happen in a network of thousands of hydrophones, AMFIP II features a built-in diagnostic system that enables the technicians to quickly pinpoint the trouble, right down to the dead hydrophone or defective integrated circuit. A major advantage of the system is that sensors can be monitored and diagnostics can be performed aboard ship, a luxury much appreciated by technicians who would otherwise have to hoist arrays of a multitude of sensors and instrument cylinders aboard to troubleshoot. "It takes an incredible amount of computer power just to do the testing," Andrews says, but it's obviously worth it. The system also allows operators to monitor system degradation in the circuitry that is immersed in the salty, high-pressure environment.

Signal Processing: On the Beam

The beamforming, or signal processing, capabilities of AMFIP II produces the high-quality measurements and acoustic images of the submarines. The image beams are computed in real time for up to 35 frequency bands simultaneously. The data from the hydrophones bobbing in Exuma Sound pours into the Hayes at a mind-boggling rate of 165 megabytes per second. I&C Division researchers Eva B. Freer and Bill Zuehzow have been instrumental in developing the algorithms and structure that bring all of the data into a usable configuration. "To get a sense of how much information that is, a 3.5-in. floppy disk holds 1.4 megabytes," Freer says. "That's 115 full high-density floppy disks per second for each array"

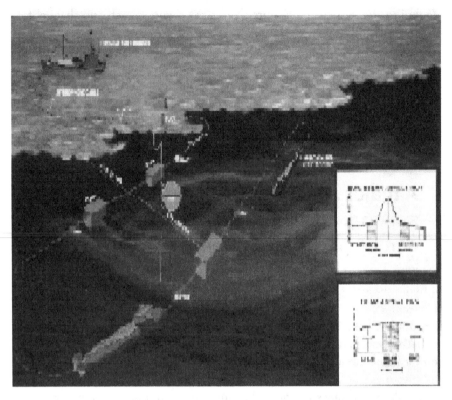

The Navy's nuclear submarines run through a course to be monitored by the AMFIP II system's sensors. The staggering amount of data generated by the sensors is manipulated by the system's signal processors and powerful computers to produce the acoustic signature of a craft that makes almost no noise.

"The custom electronic equipment aboard ship preconditions the data using more than 40 billion fixed-point instructions per second from embedded digital signal processing chips. This preconditioning is done even before the supercomputers process the data. The ability of this system to collect a multitude of signals and increase their gain with the computer algorithms enables the system to raise the submarines' image from the ocean's background noise."

As might be imagined, handling data from thousands of instruments takes a huge amount of computing power. Devising ways to handle and analyze the amount of sensor data coming in at this rate takes very specialized programming, and that is a specialty of I&C's Real-Time Systems Group. I&C Division was home to a 40-gigaflop supercomputer--for a time the biggest at the Laboratory to perform the signal processing algorithms. The result of all of this data crunching is the image of a ship, or at least its sound image.

Building and Debugging

A sizable portion of the electronics for the AMFIP II system is housed in thick-hulled pressure vessels built to stay watertight in deep water. The system's printed circuit assemblies, which are contained in the vessels, number over a thousand.

Because part of the system operates in seawater at considerable depth, it must be well built. The hydrophone arrays feature thick-hulled cylindrical instrument packages that are custom made by the Navy and pressurized with helium, which conducts heat away from the electronics. When something does malfunction, such as a sensor going out, it can be detected and compensated for aboard the Hayes. Calibration of the system can also be done remotely.

"Close to half of our effort is for testing proving that the system is working and debugging it when it doesn't," Wetherington said. "In all, the I&C Division and subcontractors have put about 30 years of conceptual development and prototyping, 20 work-years of software development and integration, and 10 years into electronics design, integration, and tests. We've also fabricated and performed quality assurance checks on 1300 multilayer printed-circuit assemblies."

The AMFIP II system consists of components and software from seven commercial suppliers, ORNL researchers, and three subcontractor teams. Putting together a product of the complexity of AMFIP II from these diverse sources required careful systems integration. "An important aspect of this project is the teamwork," Wetherington says. "ORNL had 147 staff members who worked on the effort. Our sponsor was very supportive and worked as a member of the team. We also had top-notch support from several subcontractors, including Planning Systems, Inc.; Cray Research, Inc.; Colonial Assembly and Design; and the University of Tennessee."

The acoustic signaling technology is also being considered for medical applications such as diagnosing a malfunctioning heart.

The I&C Division's AMFIP project began in 1986; the first acoustic processing system that ORNL developed was delivered in late 1989, and AMFIP II began soon after. The meticulous and ongoing attention to detail and project planning came to fruition again in October 1994 when ORNL successfully delivered and installed the first portion of the new AMFIP II measurement technology. The remaining components were installed in July 1995. The Navy sponsors have indicated that they intend to apply the technology to other tasks throughout the fleet. The acoustic signaling technology is also being considered for medical applications such as diagnosing a malfunctioning heart through its acoustic signature.

AMFIP II evolved from a project that played to the I&C Division's strengths in instrumentation, computing, and systems integration. The team of researchers also proved themselves in adapting to new projects and identifying when to go outside the Laboratory. They identified new technologies and coordinated a complex effort that involved six large organizations and a variety of engineering disciplines from electronics to high-speed computing. The astonishing stealth of the Navy's "quiet platforms," Wetherington said, made necessary the awesome amounts of data and the electronic and computing expertise and toil involved in harnessing that data. "All of that effort has been needed to lift one analog signal, a submarine's acoustic signature, up out of the ambient noise of the ocean."

I&C Division researchers, who considered customer satisfaction as one of the most important goals of the program, count the Navy as a happy customer and the product that they have delivered as a giant step forward for their division.

Introduction to Section 3

Ray Adams

Contained in this section is a part of the legacy of E. P. Epler, who was for many years, the head of the Reactor Controls Department in the Instrumentation and Controls Division.

Papers published in this section:

Lester Oakes' recollection of the beginnings of Reactor Control at ORNL

Automation of Reactor Control and Safety Systems by L. C. Oakes

Sub criticality Measurements in an LMFBR by N. J. Ackermann

A Method for Verifying Reactivity-Feedback Time Responses In Power Reactors by D. N. Fry

The I&C Analog/Hybrid Computer Facility by R.S. Stone

The Analysis of Nuclear Reactors, According to their Noise Spectra – A Summary of Kryter and Fry's work, by Ray Adams.

Section Three

Reactor Controls Department

A Compilation by Several People

Ray Adams

Several people have contributed to this section of the book. It is a compilation that represents the legacy of E. P. Epler and others who made the measurements and control of Nuclear Reactors their career for many years. It seems somehow fitting that it has been difficult to find people to write about the high art of Nuclear Reactor controls, as Epler himself certainly did not seek the limelight and was always reticent to claim credit for the accomplishments of the Reactor Controls Department.

First, is a bit of reminiscence by an early member of the Reactor Controls Department, who became its department head and was associate director of the I&C Division, and a corporate fellow.

Chapter 14

L. C. Oakes

Early History

In 1951 there was no organized reactor control group at ORNL A small group of Engineers working on controls for the enriched reactors was housed in a large lab in the old physics building near the cafeteria. It consisted of Tom Cole, E.P. Epler, Steve Hanauer, Jim Owens, and Henry Newson. Early in 1951 Hanauer and Epler moved their operation to Idaho Falls, Idaho to assist in installing the control system for the Materials Testing Reactor (MTR) for which the low Intensity Test Reactor(LITR) was an operating prototype. This group designed the control system for the LITR and most of the concepts tried on the reactor worked as planned. However early designers were uncertain about the effect of air bubbles passing through the core with the coolant. So a fast amplidyne servo was used originally. It was later determined that high speed was not needed and the amplidyne was replaced with a more simple Kochenburger type on-off servo. Jim Owens had been the chief engineer for the Bulk Shielding Reactor (BSR) instrumentation and control system with assists from the other members of the group. Tom Cole was at that time working on plans for the Oak Ridge Research Reactor (ORR). Henry Newson had gone back to work on different problems in the Physics Division. Members of the group, somewhat detached from the Physics Division, P R Bell and ?.? Strauss developed a small electronic simulator to test different control and protection systems. Two other members of other divisions were working on related problems. Walter Jordan a member of the Physics Division was involved in the design of many of the electronic instruments, a linear amplifier and a logarithmic amplifier as examples. Joe Gundlach from the Instrument Department worked on developing ionization chambers and gamma compensated neutron chambers.

*   This era was before central air conditioning became the norm. The lab where the group was quartered required air conditioning for successful operation of the instruments so there was no shortage of visitors on hot days.

*   In 1951 the Fairchild Company decided to stop work on the Nuclear Energy for the Propulsion of Aircraft (NEPA) program. Ed Bettis was head of the small reactor I & C group working there. His group had been involved with work on the Bulk Shielding Reactor at ORNL. This group had also designed a digital computer for the shielding group at ORNI,. The computer, the Oak Ridge Automatic Calculator for Linear Equations ORACLE[17], was designed to solve large systems of simultaneous equations[18]. Work on research reactors was increasing at ORNL so Bettis and any members of his department were invited to come work at ORNL. A small group consisting of Cole, Hanauer, Epler, along with L.C. Oakes, E.R. Mann who came from NEPA and a few others were organized under the leadership of Lyle Lieber (Bettis had left the group to work on the design of new reactor concepts.) Soon after the group was formed Lieber left and the group reported to Epler thereafter.

*   As soon as the 4500 building was completed the personnel moved and became known as the Reactor Control Group. The offices were in a hallway on the second floor behind the Laboratory directors office. The group along with all the equipment had laboratory space on the first floor of the Physics Division wing of corridor H. Mann began work on an analog computer for the ANP program. It was necessary to design analog components from scratch since there were no commercial systems available at that time. The heart of the computer was a high gain chopper stabilized d c amplifier. It utilized a design from an RCA Review article describing an amplifier used to study frog leg nerve impulses. Several of these units were fabricated by the Instrument Department and all performed well. Many groups unrelated to reactors used the analog computer, for example the group studying weather problems (Frank Gifford's group). Simulations were not universally accepted in the early days, some saying it looked more like electronic witchcraft. However it proved to be very useful in establishing the performance of proposed control and safety systems.

*   The concept of redundancy was established by Fermi's group that designed the Graphite Reactor The several rods there could be driven in by gravity, electric motors or hydraulics. Epler and his colleagues, Hanauer, Newson and later Mann embraced the redundancy idea but also promoted the concept of complete isolation between control and safety systems. This extended to using different basic concepts where possible such as pneumatic for control and electronic for safety systems. The rationale for this was that it is unlikely that devices based on different principles would experience random failures at the same time. It would also greatly lessen the probability of a common mode failure that would disable both systems. They also wanted system redundancy in both control and safety systems. To this end they promoted multiple servos to control the power level. This technique cannot be done with two units but is easily accomplished with three. To perform properly a servo system must move control rods at a much faster rate than is allowed for gross shim rods, to assure that a runaway control rod will not induce a damaging power excursion, the rod worth is limited to an amount that is harmless if inserted in zero time. These concepts of redundancy and separation were widely used throughout the USA and elsewhere as many organizations interested in reactors sent personnel to the controls group to study control and safety system design

*   An interesting side bar here. The NEPA group had designed a servo for the BSR. It consisted of contra-rotating shafts driving a gearbox to which the control rod was connected. Each shaft was connected to a motor through a magnetic clutch the motors were kept running at normal operating speed. The magnetic clutches were energized by error signals from the power demand system. The clutches were filled with a mixture of oil and iron filings. The concept worked very well until the iron filings were plated out on the outside of the clutch housing by centrifugal forces.. The designers thought that perhaps small ball bearings might perform better. So they contacted a ball bearing manufacturer who readily agreed to produce the required bearings. The only catch was that the price was astronomical because the factory would need to be shut down to convert to the small size bearings. That ended the novel servo idea.

Editor's Note:

Reprints from Nuclear Safety

There was for many years a publication that represented several aspects of the best of technology in this area. The editor of that journal, Edward Hagen, was himself a member of the I&C Division and many (if not most) of the articles were written by members of the I&C Division. That journal is now defunct, so I have asked R. S. Stone who for many years was a keen observer of the Nuclear Safety scene to choose several of the articles from *Nuclear Safety* for republishing. Following are three of the articles selected by Bob Stone.

Chapter 15

Automation of Reactor Control and Safety Systems at ORNL

By L. C. Oakes, Reprinted from
Nuclear Safety, Vol 11 Number 2

Abstract: Reactor safety functions requiring fast response are automated, but objections arise when some traditionally manual operations are proposed for automation. In a reactor control system, perhaps more than in any other, the operator is dependent on sensory instrumentation for determining the operating state. Comparisons of reliability, operability, economics, and safety for automatic vs. manual control are about the same if similar performance is expected. These are exemplified by discussions of automatic startups and shimming. Operator justification is even now somewhat philosophical, and results of current research and development in computer control and diagnostic techniques will further tend to reduce his role. A greater amount of automation than now used would lead to improved safety and operability.

There has never been objection to automating functions requiring speed of response beyond the capability of human senses; examples of such functions are fast safeties and servos. However, there seems to be unwarranted objection to including some traditionally manual operations that are potentially less difficult to automate. Perhaps there is a psychological reason for this attitude in the argument that an automatic system does not have the same capacity to deliberate the advisability of continuing a control sequence as does an operator. A closer look at this line of reasoning, however, leads to a different conclusion. In a reactor control system, perhaps more than in any other, the operator is dependent on sensory instrumentation for determining the operating state.

When automation of control operations is pursued further, three possible modes of operation may be recognized: (1) fully manual operation wherein the operator makes all manipulations manually with no automatic "inhibits" and his actions are guided by visual indications from sensory instrumentation; (2) manual operation wherein the reactor is maneuvered manually subject to restrictions imposed by interlocks from the instrumentation, such as rate of change of flux or power; and (3) fully automatic operation during which the operator plays no part, except surveillance, but he has veto power over the automatic systems and he selects the mode of operation and the operating power level–all other functions are performed by sensory instrumentation. In all three modes, operation proceeds or stops on the basis of information obtained mainly from the instrumentation system. Except for information on events too trivial, or whose occurrence is too improbable, to instrument, the operator has no information not presented to him through the instrumentation system. One may then ask: What differences are there between the criteria for manual systems and those for automatic systems?

REQUIREMENTS OF AUTOMATIC AND MANUAL SYSTEMS

It is necessary first to consider the reliability and operability requirements of instrumentation systems for automatic or manual operation. Clearly, if one is to automate a function, he must use instrumentation in whose reliability he has a high degree of confidence, but it is necessary that an operator using manual control have equally reliable instruments to prevent mis-operation through erroneous information. Thus a system functionally adequate for manual operation should also be adequate for automatic operation and vice versa. The requirements are about the same in both cases if comparable performance is expected.

Next, the cost of an automatic system is not much greater than that of a manual system. In a properly designed system, the operator is a gate through which information must flow. He may therefore be replaced by a suitable "and/or" logic gate.

Finally, reactor safety is an area most often thought to be jeopardized by automation, in particular when considering automatic startup. Among other things, the reactor safety system must be designed to cope with any conceivable mis-operation by the operator, the reactor control system, or the power-regulating system. Aside from assuring that a neutron source is present when a reactor is started, the startup channel serves no safety function. Therefore, after the presence of a source has been verified either by an operator or by a suitable automatic check of startup conditions, the worst that can result from startup-system mis-operation, either automatic or manual, is to produce an unwanted scram. The safety system should always be capable of handling a startup accident. The startup accident is, in fact, trivial in many cases, such as in the High Flux Isotope Reactor and other similar reactors, and the safety system is often capable of handling excursions an order of magnitude more severe. Furthermore, since an automatic startup system generally makes use of the same sensor instrumentation and the same final control elements as those used for manual startup, such as relays and motors, the probability of having an unwanted scram is about the same in both cases.

AUTOMATED REACTOR CONTROL FUNCTIONS

Although the fast safeties and servo systems have usually been automated, the concept of an automatic startup system evolved as reactors grew from novel to commonplace. The first crude automatic start was made on the Low Intensity Test Reactor (LITR) at the Oak Ridge National Laboratory (ORNL) in the late 1940's. A designer suggested, and demonstrated, that the LITR could be safely brought to power from source level by holding the manual rod-withdrawal switch closed while rod withdrawal was inhibited when the reactor period, as indicated by the log N, became shorter than 20 or 30 sec. The LITR lacked sufficient redundancy to permit that mode of operation to be incorporated, but the demonstration did suggest that routine automatic startups were feasible. The first system to incorporate automatic startup as part of the design was the Geneva Conference reactor built in 1955. Here the period information was derived from redundant counting channels. It worked well in principle, but, because of the operating environment, the reactor had only enough excess reactivity to go to full power; therefore this embodiment did not demonstrate feasibility in a more conventional application where potentially dangerous loaded excess reactivity is employed. Following construction of the Geneva Conference reactor, a system employing a single counter channel was installed in the Bulk Shielding Reactor (BSR) at ORNL. The system again worked satisfactorily, but the counter channel employed had a range of only 5 decades, after which the detector had to be moved. During detector movement and for some time following the relocation, the period information was erroneous. Therefore all rod motion had to be inhibited on reaching the highest allowable count rate and during and immediately following detector repositioning. This deficiency, along with others, fostered the development of a wide-range counting channel[1] with which the period could be determined continuously over the full range of the reactor. Furthermore, once having come on scale, the channel covered continuously the remaining range of the reactor, a principle yet very much adhered to at ORNL for any channel of instrumentation.

Starting with the Graphite Reactor and progressing to the HFIR, each new reactor built at ORNL has been equipped with more automatic systems.[2] The Graphite Reactor employed an automatic fast scram system, and the manual power-level control was replaced with a servo controller in 1953. The HFIR has the greatest number of automatic systems,[3] which include the fast scrams, the power-level controller, the system for startup from source level to sensible power, and a system for process-system startup. The process-system startup in the HFIR was automated to expedite a fast return to full power level following programmed power reductions during short-term a-c power outages. The restart sequence of the primary and secondary pumps and the primary pressurizer pumps was intricate, and slight misoperation could cause an unwanted scram at a time when xenon poisoning could prevent a restart. In fact, with the exception of automatic shimming of the control rods to compensate for reactivity decreases, the HFIR could run through a complete 20 day cycle with no assistance from human operators. Automatic shim-rod withdrawal had not thus far been incorporated in the HFIR for technical and economic reasons. Work is now under way in this area, however.[4]

Automatic shimming is done either by inserting or withdrawing the shim rods when the regulating rod reaches the end of its allotted stroke. Automatic shim-rod insertion has always been permitted in control systems at ORNL. However, automatic shim-rod withdrawal has not thus far been permitted on the basis that it violates, in a rather subtle manner, the criterion that the safety system must be able to cope with any misoperation of the servo system, including the insertion of all its reactivity as a step. It is easy to see that, if the shim rods were withdrawn when the regulating rod reached the withdraw limit and shim-rod withdrawal was subject to no other restrictions, the shim-rod withdrawal might compensate for large amounts of anomalous slowly varying negative reactivity. If the effect that produced the reactivity anomaly suddenly reverted to its original state, a large positive reactivity effect that was far in excess of the total regulating-rod worth would be impressed on the core, with possibly disastrous results. Cases of reactivity anomalies have been observed at ORNL, but they were detected and corrected before damage could result.

To make an intelligent readjustment of the regulating-rod position by withdrawing shim rods requires considerable knowledge of the core conditions. Lacking sufficient information via the instrumentation system, the shimming operation has been left to the discretion of the operator, who all too often has very little information with which to make a decision. The designers have, in fact, been accused of leaving a problem with the operator that they, the designers, cannot solve. If the reactor is operating at a steady power level and all fission-product poisons are in equilibrium, it is not difficult to shim either automatically or manually. However, correct shimming in the face of varying temperatures, power levels, and fission product poisons is difficult, if not impossible, without the aid of a computer to calculate continuously the many variables that change the reactivity. Generally an alarm sounds when the regulating rod nears the end of its stroke to alert the operator that shimming is needed. This causes many operators to become more preoccupied with preventing the alarm than weighing the need for shimming. Furthermore, we have observed a tendency on the part of operators to become conditioned to an alarm. The operator therefore reacts in accordance with an established pattern and does the same thing irrespective of the prevailing conditions. It is unreasonable to expect humans to do otherwise. In any event, one is little worse off by permitting the regulating rod to withdraw the shim rod automatically.

ROLE OF THE OPERATOR WITH AUTOMATED SYSTEMS

Having automated the fast scrams, the regulating system, the startup system, and many of the larger process systems, one may ask: What is the role of the operator? It is easy to list tasks for which he is not optimum but more difficult to find ones for which he can improve system safety or operability. It is not an optimum role for the operator to acknowledge to a reactor control system that he agrees with the instrumentation if he has only information from those instruments. If, however, by observing a greater number of channels or systems, he can deduce trends or impending failures not possible from observing a single channel alone, he can contribute to safe orderly operation. However, to achieve this, the operator must be trained in depth on the reactor and its instrumentation and control system. It is the opinion of many that the operator can best function in this capacity if the maximum amount of automation is employed to free him to concentrate on overall system behavior.

A less definable function of the operator, but one which is generally considered vital, is that of being on hand to deal with situations which, by their nature, cannot be expeditiously automated. In this category are unexpected events, functions that cannot be economically justified, or functions that cannot be included for technical reasons.

CONCLUSIONS

All automatic control and safety systems used at ORNL have performed well and have gained almost unanimous acceptance. There have been no excursions or mis-operations requiring scram action that were induced by the automatic systems.

Development work now under way in the field of computer control and diagnostic techniques will probably further reduce the role of the operator. What constitutes an optimum balance between automation and manual operation of a reactor has been the subject of much discussion at ORNL, but no definitive conclusions have been reached. It is undoubtedly a function of the reactor type, and the viewpoints held at ORNL are strongly influenced by the research type of reactor. However, the mode of operation of a research reactor is not vastly different from that of a power reactor, since both reactors have infrequent startups followed by long periods of steady power-level operation. In my opinion a greater amount of automation than now used throughout the industry would lead to improved safely and operability.

REFERENCES

1. R. E. Wintenberg and J. L. Anderson, A Ten Decade Reactor Instrumentation Channel, *Trans. Amer. Nucl. Soc.*, 3(2): 454-455 (1960).

2. J. L. Anderson and E. P. Epler, Automation in Reactor

Systems, paper presented at 1960 Nuclear Congress, New, York, N. Y., Apr. 4-7, 1960 (unpublished). ·

3. F. T. Binford and E. N. Cramer (Eds.), The High Flux Isotope Reactor—A Functional Description, USAEC Report ORNL-3572, Rev. 2, Oak Ridge National Laboratory, June 1968.

4. J. B. Bullock and H. P. Danforth, The Application of an On-Linc Digital Computer to the Control System of the High Flux Isotope Reactor (HFIR), *IEEE (Inst. Elec. Electron. Eng.) Trans. Nucl. Sci*, NS-16(1): 222-226 (February 1969).

Chapter 16

Sub criticality Measurement in an LMFBR

By Norbert J. Ackermann, Jr.
Reprinted from *Nuclear Safety*, Vol 12 Number 6

Abstract: Reliable knowledge of the sub criticality state of a nuclear reactor at all times during shutdown, coupled with proper administrative control, should preclude the possibility of that reactor accidentally becoming critical or supercritical This review of the state of the art of sub criticality measurement in the LMFBR gives particular attention to four measurement techniques: neutron-source multiplication, neutron-noise analysis, inverse kinetics and pulsed neutrons. It is concluded that the neutron-source multiplication technique is the only method applicable for measuring the sub-criticality in an LMFBR over the full range of shutdown. Present sub-criticality-measurement development programs are reviewed, and future applications are discussed

The accurate determination of the reactivity of a sub critical nuclear reactor has been of interest to the reactor community for many years. The reasons for determining the sub criticality of a reactor are discussed in this article, and various sub criticality-measurement techniques are reviewed and evaluated for practical on-line application in a liquid-metal-cooled fast breeder reactor (LMFBR).

Reliable knowledge of the sub criticality state of a nuclear reactor at all times during shutdown, coupled with proper administrative control, is needed to preclude the possibility of accidental criticality or super criticality during reactor maintenance and refueling operations. The probability of accidentally increasing the reactivity of a reactor is a maximum during maintenance and refueling operations, and the consequences of having an accident during these operations would be severe because many of the accident-limiting mechanisms of the reactor are inoperative at such times due to interference with operation (e.g., the reactor pressure vessel may be open for refueling).

Present schemes of reactor maintenance and refueling are generally formulated under a set of very conservative and restrictive ground rules to protect against a reactivity accident in a sub-critical reactor. These rules are necessary to guard against a reactivity accident since an accurate and reliable estimation of the exact sub-criticality state of the reactor has thus far been judged to be reasonably unobtainable. If accurate and reliable knowledge of the reactor's sub-criticality state were available from some measurement system, some of these time-consuming restrictions on reactor maintenance and refueling could be relaxed with no decrease in plant safety but with considerable economic saving. Also, since sub-criticality cannot be measured accurately at all times during shutdown, excessive shutdown margins are required as a safety precaution. This escalates reactor costs because more control rods must be provided and space must be made available for them in the reactor core. The extra shutdown margin also increases reactor startup time. However, if an alternate and independent shutdown system is provided, as is required by U.S. LMFBR design, then the concern for excessive shutdown margins becomes moot.

A sub-criticality-measurement system would be especially helpful in establishing, through a direct measurement. that the reactor is fully shut down after a scram, scheduled or unscheduled. This information is certainly needed, for example, following a reactor accident such as that in the SL-I and an incident such as that in the Fermi reactor. In such cases it is necessary to determine, as quickly and as accurately as possible, the precise shutdown state of the reactor in order to be able to guard against any further incidents.

During the late 1950s and early 1960s, numerous techniques for measuring the sub-criticality of a reactor were proposed and examined in search of a method to be used in the emerging U. S. light-water power-reactor systems. For several reasons, however, this flurry of research activity was unsuccessful in developing a practical and reliable sub-criticality-monitoring system and having it incorporated into the light-water reactor (LWR) plants of the U. S. nuclear power industry. The search continued, nevertheless, with new techniques being proposed and old techniques being refined and modified in the hope of eventually developing a system that would be useful.

With the expansion of interest in the sodium-cooled fast breeder reactor that was brought about by the establishment of the AEC LMFBR program, it was only natural that the question would arise of whether a sub-criticality-monitoring system could be developed for use in an LMFBR. In the LMFBR program plan[1], the requirements that were established for the sub-criticality-measurement system were that it should be able to measure the reactivity in the range from critical to 15 dollars sub-critical with an accuracy of 5% and below 15 dollars with an accuracy of 20%. An alternative criterion that has been suggested for the sub-criticality-measurement system is that it should be able to measure the sub-criticality at full shutdown (~30 dollars) to an accuracy of 20% and that while the reactor is fully shut down the measurement system should be able to measure a 4-dollar change in reactivity to an accuracy of 25%. The incentive for development of a sub-criticality- measurement system for an LMFBR was the realization that the severity of a shutdown reactivity accident could be potentially much greater in an LMFBR than in an LWR because of the possibility of an uncontainable sodium fire.

This article is a summary of a state-of-the-art report which was prepared at the request of the AEC Division of Reactor Development and Technology and which constituted the first step in the LMFBR program plan for developing an on-line sub-criticality measurement system for an LMFBR[2]

MEASUREMENT CONDITIONS

A realistic evaluation of a sub-criticality measurement technique for application to an LMFBR must include consideration of the measurement conditions that would be encountered. All sub-criticality measurement techniques of interest are based on sensing variations in the neutron population of the reactor during shutdown; this is called source-range detection or low-level-flux monitoring.

The choice of a low-level-flux monitoring system for a reactor necessarily involves compromise. For example, when sub-criticality is to be measured, it is desirable that the low-level-flux monitor have as high a neutron-detection efficiency (i.e., neutrons detected per reactor fission) as possible. However, placement of the detector in the reactor is limited by space availability and refueling-interference problems to the extent that detection efficiency is low. The detector is usually as far from the core region as it can possibly be and) yet retain the ability to monitor the shutdown flux level of the reactor. Furthermore, in an LMFBR, as in all power reactors, low-level neutron detection is complicated by the presence of a high gamma. background level from the large fission-product inventory. A fission counter that uses pulse-height discrimination techniques is the only type of neutron detector that has demonstrated capability for adequately detecting neutrons in such a high gamma background,[3,4] Compared with a boron or helium ionization chamber. however, a fission counter imposes a rather severe loss of neutron sensitivity.

A further complication in the neutron detection problem is a general reduction in neutron sensitivity of all detectors due to the harder neutron spectrum in fast reactor compared with that of a thermal reactor and the decreasing detection cross section with increasing neutron energy. Calculations of the neutron. detection efficiency of a fission counter (coated with 2 g of ^{235}U, as is typical of low-level-flux monitors) in the Fast Test Reactor (FTR) of the Fast Flux Test Facility (FFTF) shows that the neutron-detection efficiency would be on the order of 10^{-6} in the reactor core and inner reflector and would decrease to 10^{-8} and lower at the outside of the shield.[5] Consequently an evaluation of sub-criticality measurement techniques for practical application in LMFBRs must give consideration to the fact that the neutron-detection efficiency of the low-level-flux monitors will be in the range 10^{-6} to 10^{-8}.

Among other factors to be considered is that the LMFBR has a large, distributed, inherent neutron source due to spontaneous fissioning and to the (a,n) reaction of the oxide fuel. In the FTR, this source is expected to be about 2×10^8 neutrons/sec. The prompt-neutron kinetics of the LMFBR will occur on a much shorter time scale relative to a thermal reactor because of the much shorter neutron lifetime in an LMFBR ($\sim 5 \times 10^{-7}$ sec vs. $\sim 10^{-4}$ sec). The faster response of the LMFBR imposes a requirement of much faster instrumentation system response for the techniques based on prompt-neutron kinetics.

EVALUATION OF SUBCRITICALITY MEASUREMENT TECHNIQUES

Examination of the various measurement techniques indicated only four techniques that are promising from a practical standpoint and therefore warrant consideration: neutron-source multiplication, pulsed neutrons, inverse kinetics, and neutron-noise analysis. The source multiplication technique is a statics method wherein the sub-criticality is inferred from the steady-state neutron level. The other three techniques are kinetics methods in which sub-criticality is inferred from the analysis of time variations in the neutron level. Each of the four sub-criticality-measurement techniques is evaluated individually in the following for application to an LMFBR. The basic concept of each measurement technique is reviewed, and the major advantages and disadvantages for use in an LMFBR are presented. Also, the most pertinent application to date of each technique in terms of use in an LMFBR is cited.

The Neutron-Source Multiplication Technique

The well-known, time-honored neutron-source multiplication technique has been applied since the very beginning of nuclear reactors. This technique is used to monitor all reactor loadings to critical and, at least in a qualitative way, all subsequent fuel reloading. Serber presented the first definitive treatise on the method,[6] which is based on the general concept that a neutron source emitting S_0 neutrons/sec in a reactor with multiplication constant K results in a multiplied source s that is the sum of all the neutron generations, or

$$s = So + kSo + k^2So + \ldots + k^nSo + \ldots = So/(1-k) \quad (1)$$

The observed steady-state detector count rate, CR, in a sub-critical reactor is related to the sub-criticality in dollar units, $, by

$$CR = WSo/ (v\beta\ [-\$]) \quad \textbf{(2)}$$

where W is neutron-detection efficiency (detections per reactor fission), v is the average number of neutrons produced per fission, and β is the effective delayed-neutron fraction.

The neutron-source multiplication Technique is used at least in a qualitative way in all reactor shutdowns through the monitoring of the steady-state neutron-detector responses. If an unexpected increase in detector count rate is observed, the operator is alerted to the fact that an unexpected increase in k or $ may have occurred; this follows from Eq. 2. Many times Eq. 2 is used to infer a quantitative estimate of $; however, this requires that all the calibration factors wS_0/v_rB be determined. This can be done by calculation but usually is done by a calibration measurement. Changes in the steady-state count rate are then interpreted as changes in $ by assuming that the calibration factor remains constant.

Equations I and 2 are based on the assumption that the neutron-flux distribution is described by the fundamental mode. This assumption is valid near critical but is questionable as the reactor is made far sub-critical because of harmonic-mode contamination of the flux distribution. The degree of validity depends on the position of the neutron source, the detector location, and the means employed in shutting down the reactor. Paxton and Keepin have described the great lengths to which experimenters have gone in choosing proper detector and source locations to minimize higher mode contamination in the source multiplication measurement.[7] In a power reactor, however, it is not possible to move sources and detectors at will. Because of this, Walter developed a method that accounts for the harmonic-mode contamination in the measurement by directly calculating the contribution to the detector response of each neutron generation, beginning with the primary source neutrons.[8] The calculation is performed by adapting the iteration scheme used in a typical reactor calculational code whereby each successive iteration in the calculation corresponds to a successive neutron generation.

In an LMEBR the neutron source [spontaneous fission and (a,n)] is uniformly distributed within each fuel zone, so harmonic-mode contamination is minimized, especially if the shutdown is achieved by uniform rod insertion. Thus Eq. 2 may be generally valid over the full range of shutdown (~30 dollars). However, changes in the calibration factor must be accounted for, particularly changes in *W*, which usually occur due to rod-shadowing effects. The changes in W can be determined by calculation of the neutron flux distribution for the reactor conditions of interest and then use of the following equation:

$$W = \frac{\int V_d \int AII \ E \ \Sigma_d(r,E) \ \phi(r,E) \ dr \ dE}{\int V_c \int AII \ E \ \Sigma_f(r,E) \ \phi(r,E) \ dr \ dE} \qquad (3)$$

where $\Sigma f(r,E)$ and $\Sigma d(r,E)$ are the macroscopic fission and detection cross sections and Vc and Vd are the core and the detector volumes, respectively. *W* is very sensitive to a local perturbation in the flux distribution since the numerator of Eq. 3 is integrated in space over the volume of the detector only, whereas the denominator is integrated over the entire reactor core.

Changes in *W* are likely to occur during refueling because the neutron source is distributed with the fuel; however, in general, these changes should be small. Monitoring of the fuel elements for inherent neutron source level before their insertion into the reactor would provide a check of source-level changes during refueling. Insertion into the reactor of material with an appreciable (γ,n) cross section would also cause a change in the neutron-source level since the gamma background level at shut down in an LMFBR is quite high. These changes in S_0 would have to be included in the interpretation of Eq. 2.

The major advantages of the source multiplication technique are its simplicity and ease of application. No extra equipment is needed since it utilizes the normal reactor low-level-flux instrumentation.

Smith has used the neutron-source multiplication technique to track the sub-critical reactivity in the Dounreay Fast Reactor (DFR) during a fuel reloading by using Eq. 2 to interpret the detector response.[9] The calibration of Eq. 2 was made by the removal of some well-calibrated fuel-bearing control rods so that the reactor was initially shut down by a small, known amount. The DFR was then fully shut down (~$20), and the sub-criticality was monitored through the use of Eq. 2 while 90 standard fuel elements and several special elements were reloaded. The net change in reactivity resulting from the large fuel exchange, as determined by the neutron-source multiplication measurement, was found to be in error from the actual value by only 0.5 dollar.

A variation of the neutron-source multiplication technique, called the asymmetric-source technique, has been proposed by Walter and Henry.[10]

They have shown that the sub-criticality of a reactor may be inferred from the measured flux asymmetry that results from the insertion of a neutron source into an asymmetric position in a sub-critical reactor. As the degree of sub-criticality increases, the spatial neutron flux distribution becomes more and more peaked about the asymmetrically located source position. From comparisons of measured flux asymmetries with calculated flux asymmetries for various reactor sub-criticalities, the degree of sub-criticality may be inferred.

The asymmetric-source technique has the same major disadvantage as the neutron-source multiplication technique; namely, calculation of neutron detection efficiency changes caused by control rod insertions. However, in the case of the asymmetric-source technique, these calculations are even more difficult because they must include the asymmetrically located neutron source. Furthermore, the asymmetric. source technique requires the insertion of a much stronger neutron source than the already present inherent neutron source. This can be accomplished only through use of large quantities of ^{252}Cf. Thus the asymmetric-source technique is not deemed to be more advantageous than the neutron-source multiplication technique for measuring sub-criticality in an LMFBR.

The PulsedNeutron Technique

The pulsed-neutron technique is based on the observation of the time die-away of the neutron population following the introduction of a large-very-short-duration burst of neutrons from a pulsed neutron source placed in or near the reactor core. The time die-away of the neutron population, $n(t)$, is assumed to follow a fundamental-mode decay,[11] after the decay of higher modes, given by

$$n(t) = n_o\, e^{-\alpha t} \qquad (4)$$

where

$$\alpha = (\beta/\Lambda)(1 - \$) \qquad (5)$$

and Λ is the prompt-neutron generation time. In the pulsed-neutron measurement, the exponential decay constant α is determined from a parametric fit of Eq. 4 to the experimental data, and then α is related to $\$$ through Eq. 5. The measurement at critical gives β/Λ, which is assumed to remain constant as the reactor is made sub-critical. This assumption is valid near critical but is questionable when the reactor is far shut down. Pulsed-neutron measurements ate difficult to make at or very near critical, so β/Λ is usually determined from extrapolation of data obtained near critical.

Since the pulsed-neutron measurement depends on a strong perturbation to the neutron-flux level, very precise measurements of α, and thus $, can be made, in principle, in a short time. However, this perturbation excites harmonic modes (since it is introduced at a point), and, as the sub-criticality increases, the persistence of the harmonic modes relative to the desired fundamental mode increases. Consequently it becomes very difficult to observe a fundamental-mode decay in a far sub-critical reactor. Because of this, it is doubtful that the results of the pulsed-neutron technique would be interpretable over the full range of shutdown in the large, heterogeneous core of the LMFBR. The large, inherent neutron source of the LMFBR is another complicating factor in the measurement since the pulsed-neutron source would have to override this background source.

The need for a pulsed-neutron generator to operate in or near the reactor core is an even more severe limitation to the application of the pulsed-neutron technique as an on-line monitoring system in an LMFBR. Because of the hostile environment in an LMFBR core, reliable long-term operation of a pulsed source under such conditions is presently unattainable. Drift-tube-type accelerators have been suggested for removing the vulnerable parts of the generator to a less severe environment; however, drift-tube lengths are limited to about 7 ft without additional beam focusing, and this is not a sufficient distance to relieve the problem.

Modified pulsed-neutron techniques have been employed that utilize an analysis of both the prompt. and the delayed-neutron response to the source perturbation in determining $.[12-14] These techniques require neither a measurement at critical to determine β/Λ nor the assumption that β/Λ remains constant. However, the modified pulsed-neutron techniques are still subject to the limitations cited previously for the pulsed-neutron technique. Furthermore, the modified pulsed-neutron techniques are subject to even more spatial effects, which would further complicate their on-line application.[15, 16]

In addition to the pulsed-neutron source, the performance of the pulsed-neutron measurement requires additional electronic equipment for timing and logic operations, as well as a multichannel time analyzer to record the neutron-population die-away. Lehto[17] has reported the use of the pulsed-neutron technique for measuring sub-criticality in the plutonium fueled ZPR-3 assembly 54. He was able to make measurements down to approximately 5 dollars sub-critical before the combination of higher mode contamination and the high-background neutron source made the measurement impractical.

The Inverse-Kinetics Technique

The inverse-kinetics technique was formulated by Sastre, who showed that sub-criticality could be determined by solving the point reactor-kinetics equations in an inverse manner.[18] The detector count rate and the neutron-population level as functions of time are related by

$$CR(t) = (Wn(t)/(v\Lambda) \tag{6}$$

where the proportionality constant is assumed to be constant with time. The point reactor kinetics equations are

$$dn/dt = (n\beta_i/\Lambda)(\$ - 1) + \sum \lambda_i Ci + S \tag{7}$$

where the summation is from $i = 1$ *to* M
and where Ci = precursor population for group i

$$dCi/dt = (n\beta_i/\Lambda) - \lambda_i Ci \qquad for \ i = 1, \ldots, M \tag{8}$$

and β_i = fractional yield of precursors of type i per fission
  neutron

$$\beta = \sum \beta_i \quad \text{(summation from } i = 1 \text{ to M)} \tag{8A}$$

λ_i = decay constant for precursor group i

S = neutron-source strength

The inverse-kinetics technique is usually implemented by relating a neutron detector's time-dependent response following a change in reactivity (e.g., a control-rod drop) to the neutron population by use of Eq. 6 and then solving Eqs. 7 and 8 for $\$$, with the n(t) from Eq. 6 as input. The solution of Eqs. 7 and 8 is complicated for an LMFBR by the existence of the large, inherent, background neutron source, which is an unknown in Eq. 7. The source term must be determined by an iterative trial-and-error procedure;

that is, by searching for the source term that gives a constant $ solution to Eq. 7 after the reactivity change is known to have been completed.

The inverse-kinetics technique can be relatively easily implemented in an LMFBR by use of the regular low-level-flux instrumentation and the reactor control rods. The inverse solution to the point kinetics equations can be effected through use of analog circuitry[18] or through use of a digital computer,[19] with the latter being the more versatile approach.

A major disadvantage of the inverse-kinetics technique is the loss of dynamic range in the measurement as the reactor becomes far shut down. The neutron level changes very little when a control rod is dropped in a reactor that is far shut down, and thus the neutron-transient data from which the reactivity is to be extracted are sparse. Since low neutron-detection efficiencies are to be expected in LMFBRs, as discussed previously, the CR(t) data for an inverse-kinetics measurement will have large statistical uncertainties. Consequently the sparse transient data coupled with large statistical uncertainties will result in highly imprecise estimates of $ for rod drops in a reactor that is far shut down.

A further complication in the measurement is the assumption of point reactor-model dependence. As the reactor becomes more sub-critical, this assumption becomes questionable. The use of large amounts of reactivity in the rod drop to improve precision in the measurement of $ will accentuate non point-model effects. Furthermore, the assumption of constant neutron-detection efficiency during the measurement may be invalidated if the rod drop entails a large amount of reactivity.

When fission counters are being used for inverse-kinetics measurements, special caution must be taken to account for counting losses. A change in counting loss is, in effect, a neutron-detection efficiency change, and this must be accounted for in the measurement if it occurs. This will generally be the case when rod drops are made from critical.

Carpenter has reported measurements made with the inverse-kinetics technique down to 7 dollars sub-critical in the AETR critical assembly.[19] Improvements in the inverse-kinetics-solution algorithm by Cohn[20] have made measurements down to approximately 10 dollars sub-critical possible in the plutonium-fueled ZPR-9 with an FTR-3 loading.

The Neutron-Noise-Analysis Technique

The determination of the sub-criticality of a reactor from the analysis of the randomly occurring fluctuations (noise) of the neutron density is well established.[22,23] Borgwaldt and Stegemann presented a generalized theory for neutron-noise measurements in the time domain (correlation function) and in the frequency domain (power spectral density) for both one-detector and two-detector experiments.[24] This theory is based on the point reactor-kinetics model. In the neutron-noise measurements, the sub-criticality is inferred from the prompt-neutron fundamental-mode decay constant cr in the same way as for the pulsed-neutron technique (see Eq. 5). The time-domain noise measurement yields α as the exponential decay constant of the reactor impulse-response function, whereas the frequency-domain measurement yields α as the break frequency of the reactor transfer function. These two measurements are simply Fourier-transform pairs. Edelmann has shown that the type of time-domain noise measurement that is classed as the Rossi-α measurement should be considered as an inefficient form of the more general correlation function measurement.[23]

The primary advantage of neutron-noise methods for measuring sub-criticality is that these methods are nonperturbing to the system. No special apparatus is needed in the reactor, and no reactor transients are required. Since the noise techniques require no perturbation to the reactor, the assumption of point reactor kinetics is probably valid over a larger range of sub-criticalities than for measurement techniques that require large perturbations, such as inverse kinetics and pulsed neutron. However, since the noise techniques rely on the transient information of the inherent stochastic fluctuations of the neutron density, they require a long measurement time or a high neutron-detection efficiency[26] or both to attain adequate statistical precision in a given measurement. This is the major limitation to the noise-measurement techniques.

Even though no special equipment is needed in the reactor to perform the noise measurement, a special noise-data analyzer is required to generate the correlation function or the power spectral density. Both digital[27,28] and analog[26,29] noise-data-analysis methods have been developed, but at present the analog frequency-domain measurement appears to be the most practical for an on-line LMFBR system.

Seifritz and Stegemann have reported measurements of sub-criticality down to 4.3 dollars on the coupled fast-thermal Stark reactor with an on-line meter based on the two-detector cross-power spectral-density method (2D-CPSD).[30] An evaluation of sub-criticality measurements in an LMFBR with the 2D(PSD method showed that satisfactory measurements down to approximately 3 dollars sub-critical could be made under practical conditions, with the neutron-detection efficiency being the limiting factor.[5]

Reported[31,32] variations of the measurement of sub-criticality by neutron-noise analysis are based on just the prompt-neutron multiplication information contained in the low-frequency portion of the neutron power spectrum below the prompt-neutron break frequency instead of on an analysis of prompt-neutron time decay. These measurements, unlike the conventional noise measurements, are systematically dependent on the neutron-detection efficiency,[32,33] and as such must account for neutron-detection efficiency changes in a similar way as for the previously mentioned neutron-source multiplication method. However, they are not systematically dependent on β/Λ as the conventional noise methods are. Consequently the combined use of these noise measurements would provide a complementary check of the sub-criticality. measurement dependence on β/Λ and W.[22]

PRESENT DEVELOPMENT EFFORTS AND FUTURE APPLICATIONS

At present, there is a program in progress at the Oak Ridge National Laboratory for the development of on-line sub-criticality-measurement techniques for use .n LMFBRs. The use of the neutron-source multiplication technique is being investigated over the full range of sub-criticalities in an LMFBR after calibration near critical by either inverse-kinetics or neutron-noise analysis. As part of this program, extensive experimental evaluations are so be carried out in the critical mock up of the FTR in the ZPR 9 at Argonne National Laboratory and also in the Southwest Experimental Fast Oxide Reactor (SEFOR) of the General Electric Company, with prime consideration being given to the engineering problems involved in LMFBR application.

Sub-criticality measurements made with primarily the inverse-kinetics and neutron-noise methods are being employed at Argonne in its LMFBR critical experiments program to provide LMFBR-design data. Valuable measurement-technique development is a by-product of this work, although it is not directed toward specific on-line application in an LMFBR. Also, some development work is being performed at the Hanford Engineering Development Laboratory (HEDL) and the University of Washington on the polarity-spectrum coherence method for measuring sub-criticality in LMFBRs.

The designs of the FTR of the FFTF call for the measurement of sub-criticality over the full range of shutdown. At present, there are plans to use the neutron-source multiplication technique, with neutron noise and inverse kinetics being considered for calibration.[34] The designs of the LMFBR demonstration plants of all three AEC project-definition-phase contractors also specify that sub-criticality be measured over the full range of shutdown. The techniques to be used in the demonstration plants are still undecided.

SUMMARY

The neutron source multiplication technique is the only method that appears to be applicable for measuring the sub-criticality in an LMFBR over the full range of shutdown (critical to 30 dollars sub-critical). However, it must be calibrated at a known sub-criticality and is subject to inevitable changes in its calibration factor, which must be continually corrected. The three kinetics techniques (pulsed neutron, inverse kinetics, and neutron-noise analysis) considered in this article can cover only a limited range of shutdown for the various reasons stated. However, they could be used for calibrating the neutron-source multiplication technique near critical (preference is given to the noise or the inverse-kinetics techniques, since they do not require a pulsed-neutron source in the reactor). The calibration of the source multiplication technique could be provided also by the insertion of previously well-calibrated control rods, but this requires the questionable assumption that the rod calibration has not changed, whereas the noise or the inverse-kinetics measurement would involve a direct calibration of the neutron-source multiplication technique.

REFERENCES

1. Argonne National Laboratory, Liquid Metal Fest Breeder Reactor Program Plan, IJSAEC Reports WASH-1104 and WASH-1109, August 1968.

2. N. J. Ackermann, Jr, State-of-the-Art Report on Sub-criticality Measurements in LMFBRs, USAEC Report ORNL-TM-2917, Oak Ridge National Laboratory (to be published).

3. C. N. Jackson, Jr., and N. C. Hoitink, High Temperature and Gamma Effects on Counting-Type Neutron Flux Detectors, *IEEE (Inst. Elec. Electron. Eng.), Trans. Nucl. Sci.* 17(1,Pt. 1): 369-375 (1970).

4. D. P. Roux, J. T. De Lorenzo, and C. W. Ricker, A Neutron Detection System for Operation in very High Gamma Fields, *Nucl. Appl. Technol.* 9: 736-743 (1970).

5. A. R Buhl and. N. J. Ackermann, Jr., Precision of Shutdown Margin Measurements Using the Two Frequency Reactor Noise Technique in an LMFBR, *IEEE (Inst. Elec. Electron. Eng.), Trans. Nucl. Sci,* 18(1): 430-434 (1971).

6. R Serber, The Definition of "Neutron Multiplication," USAEC Report LA-335 (Rev.). Los Alamos Scientific Laboratory, July 25,1945.

7. H. C Paxton and G. R Keepin, Criticality, Chap. 5 in the *Technology of Nuclear Reactor Safety, Vol. 1, Reactor Physics and Control.* T. J. Thompson and J. G. Beckerly (Eds.). pp. 244-284, The M.I.T. Press, Cambridge, Mass., 1964.

8. J. F. Walter, Detector Response to the Sub-critical Reactor, *Nucl. Appl.* 3: 271-274 (1967).

9. D. C Smith, Factors Involved in Time To Start Up Fast Reactors, Including Measurement of Sub-critical Reactivity, USAEC Report ANL-6792, Argonne National Laboratory, pp. 663-678,1963.

10. J. F. Walter and A F. Henry, The Asymmetric Source Method of Measuring Reactor Shutdown, Nucl. Sci. Eng., 32: 332-341 (1968).

11. B. E. Simmons and J. S. King, A Pulsed Neutron Technique for Reactivity Determination, *Nucl. Sci Eng.*, 3: 595-608 (1958).

12. N. G. Sjostrand, Measurement on a Sub-critical Reactor Using a Pulsed Neutron Source, *Ark. Fys*, 11: 233-246 (1956).

13. E. Garelis and J. L. Russell, Ir., Theory of Pulsed Neutron Source Measurements, *Nucl. Sci. Eng.*, 16: 263-270 (1963).

14. T. Gozani, A Modified Procedure for the Evaluation of Pulsed Source Experiments in Sub-critical Reactors, *Nukleonik.* 4: 348-349 (1962).

15. M. Becker and K. S. Quisenberry, Spatial Dependence of Pulsed Neutron Reactivity Measurements, in Neutron Dynamics and Control, Tucson, Ariz., April 5-7, 1965, D. L. Hetrick and L. E. Weaver (coordinators), *AEC Symposium Series*, No.7 (CONF-650413), pp.245-276, May 1966.

16. C. A. Preskitt et al., Interpretation of Pulsed-Source Experiments in the Peach Bottom HTGR, *Nucl. Sci Eng.*, 29: 283-295 (1967).

17. Argonne National Laboratory, Reactor Development Program Progress Report, November 1968. USAEC Report ANL-7518,1968.

18. C. A. Sastre, The Measurement of Reactivity, *Nucl. Sci. Eng.*, 8: 443-447 (1960).

19. S. G. Carpenter, Reactivity Measurements in the Advanced Epithermal Thorium Reactor (AETR) Critical Experiments *Nucl. Sci. Eng.*, 21: 429-440 (1965).

20. C. E. Cohn and J. J. Kaganove, Digital Method for Control Rod Calibration, *Trans Amer. Nucl. Soc.*, 5(1): 388-389 (June 1962).

21. C. E. Cohn, private communication, November 1970.

22. R. E. Uhrig (Coordinator), *Noise Analysis in Nuclear Systems*, Gainesville, Fla., November 4-6, 1963, AEC Symposium Series, No. 4. June 1964.

23. R. E. Uhrig (Coordinator), *Neutron ,Noise, Waves, and Pulse Propagation*, Gainesville, Fla., February 14-16, 1966, AEC Symposium Series, No. 9 (CONF-660206), May 1967.

24. H. Borgwaldt and D. Stegemann, A Common Theory for Neutronic Noise Analysis Experiments in Nuclear Reactors, *Nukleonik*, 7(6): 313-325 (1965).

25. M. Edelmann, New Rossi-α-Measurement Methods, USAEC Report LA-tr-68-39, translated from Report INR4/68-15 1968.

26. W. Seifritz, D. Stegemann, and W. Vth, Two-Detector Cross-Correlation Experiments in the Fast-Thermal Argonaut Reactor (STARK). in *Neutron Noise, Waves, and Pulse Propagation*, Gainesville, Fla., February 14-16, 1966, R. E. Uhrig (Coordinator), AEC Symposium Series No. 9 (CONF-660206), pp. 195-216, May 1967.

27. C. E. Cohn. Reactor Noise Studies with an On-Line Digital Computer, *Nucl. Appl.*, 6: 391-400 (1969).

28. R. C. Kryter, Application of the Fast Fourier Transform Algorithm to On-Line Reactor Diagnosis, *IEEE (Inst. Elec. Electron. Eng.)*, Trans Nucl. Sci., 16(1): 210-217 (1969).

29. T. Nomura, S. Gotoh, and K. Yamaki Reactivity Measurements by the Two-Detector Cross correlation Method and Supercritical-Reactor Noise *Analysis, Neutron Noise, Waves, and Pulse Propagation*, Gainesville, Fla., February 14-16, 1966, R. E. Uhrig (Coordinator), AEC Symposium Series, No. 9 (CONF-660206), pp. 217-246, May 1967.

30. W. Seifritz and D. Stezemann, An On-Line Reactivity Meter Based on Reactor Noise Using Two-Detector Cross Correlation, *Nucl. Appl.*, 6: 209-216 (1969).

31. W. Seifritz, The Polarity Correlation of Reactor Noise in the Frequency Domain, *Nucl. Appl. Technol.*. 7: 513-522 (1969).

32. N. J. Ackermann, Jr., and J. C. Robinson. A Noise Technique for Measuring Reactivity Independent of β/Λ, *Trans Amer. Nucl. Soc.*, 13(2): 764 (November 1970).

33. N. J. Ackermann, Jr., A. R. Buhl, and R. C. Kryter, An Analytical and Experimental Evaluation of Detection Efficiency Dependence of Sub-criticality Measurements by the Polarity Spectral Coherence Method. *Trans. Amer. Nucl. Soc.*, 14(1): 44-45 (June 1971).

34. Private communication from R. A. Bennett, WADCO Corp., Richland, Wash. 1971.

Chapter 17

A Method for Verifying Reactivity-Feedback Time Responses In Power Reactors

D. N. Fry Reprinted from , Vol 13 Number 4

Editor's Note: The following paper is included in this book, because it represents an extremely important contribution to Nuclear Reactor analysis. For additional information on this subject, the reader is referred to the work of Kryter and Fry in Chapter 19.

Abstract: A method is presented for on-line monitoring of the reactivity-feedback rime constants for determining the operating stability of large fast breeder reactors. In this method, monitoring can be performed without interfering with normal reactor operation, with negligible reactor power, disturbance, and without the need for special in-core reactivity oscillators. The method also is conducive to the detection of partial flow blockages in coolant channels in reactor core sub assemblies.

An analytical investigation was made to develop an experimental method for verifying the reactivity-feedback time constants used in the safety analysis of reactors. Knowledge of these constants is important in studies of reactor stability and safety. Hummel and Okrent[1] state that large fast breeder reactors are expected to have a positive coefficient due to sodium density change and radial fuel-pin expansion and to have negative Doppler and structural-expansion coefficients. They further state that "the relative time sequence in which temperature changes, occur may have a profound effect on the stability of reactor operation and on the role of various reactivity feedback mechanisms in accident analysis[2]. Therefore it is imperative that a method be developed which will enable measurement of the magnitude and time constants of reactivity feedback in fast reactors without interfering with normal operation and without the necessity of installing special reactivity oscillators.

If the reactor power-to-temperature transfer function, a dynamic characteristic of the core, can be measured, then the time response of the reactivity feedback from the different reactor core regions can be d e t e r mi n e d . Cross-power-spectral-density (CPSD) analysis techniques were used by Randall and Pekrul[3] to measure temperature feedback-reactivity coefficients in the SNAP-8 Experimental Reactor and by Rajagopal[4] to infer the fuel time constant in the Saxton pressurized-water reactor. Although their results demonstrated the feasibility of this technique, the method is still not used in operating power reactors to monitor for changes in the time constants because it distracts from normal operation. The analytical studies using an analog model of the High Flux Isotope Reactor (HFIR) were supported by experiments performed at the HFIR to verify the model and to demonstrate that nonperturbing tests can be conducted using existing reactor instrumentation. An analog simulation of the HFIR demonstrated that dynamic feedback information, i.e., transfer functions, can be obtained on-line with negligible reactor power disturbance and without the need for special in-core reactivity oscillators.

The HFIR[5] was chosen for study because (1) it has both positive and negative temperature coefficients, (2) an analog-computer model of the reactor has been developed by Burke[6], and (3) a technique has been developed to oscillate the control rod at different frequencies with the reactor at full power to obtain the power-to-reactivity transfer function[7].Simulated experiments using a white-noise reactivity driving function were conducted on the analog model to determine the sensitivity of the power-to-reactivity transfer function to changes in overall positive or negative temperature reactivity feedback, temperature coefficients of reactivity, and core coolant flow. The temperature-to-power transfer functions were also determined, and from these we have inferred the time constants of reactivity feedback from core, target, and control-rod regions. The sensitivity of temperature-to-power transfer functions to changes in coolant flow was also studied. Although the HFIR is a stable reactor, other reactors may not be, and the techniques developed at the HFIR should be useful for investigating their stability.

DESCRIPTION OF MODEL AND COMPUTER STUDIES

The analog-computer model of the HFIR used in these studies was originally developed for safety and control studies[6] It employed a six-delay-group nonlinear point reactor description of the neutron kinetics and a lumped-parameter method to describe the heat transfer. Five regions were considered, together with their associated temperature reactivity feedback in units of $(\Delta k/k)/$ °F:(I) fuel (-I x 10^{-5}), (2) core-water (-9.7 x 10^{-5}), (3) target coolant-water (+3.2 x 10^{-5}), (4) target annulus (+2.7x 10^{-5}), and (5) control-rod coolant water (+ 1.1 x 10^{-5}). The simulation was made in real time so that the same equipment and data handling procedures could be used for the computer studies as would be used in experiments at the reactor.

The frequency response of the analog model was obtained by supplying an external reactivity input from a white-noise generator and recording on magnetic tape the reactivity fluctuations and the resulting power and temperature fluctuations of the model. The analog tapes were digitized, and the frequency response was calculated by computing the CPSD of the input and output signals, using the fast Fourier-transform code CROSSPOW[8]. The gain was computed as:

$$G(f) = (\ |\ CPSD(f)|\ )/\ APSD\ (f) \qquad (1)$$

where APSD(f) is the power spectral density of the input driving function at frequency f. Also, the phase of the transfer functions was computed from the real and imaginary components of the CPSD(f). The input reactivity perturbations were approximately 0.1 rms cents/$\sqrt{Hz}$, and the corresponding power fluctuations were ±2 MW, about a mean power level of ~ 97 MW.

EXPERIMENTAL VERIFICATION OF HFIR ANALOG MODEL

Pseudo random reactivity perturbations were introduced into the HFIR in an experiment by Chen[9] to determine the power-to-reactivity transfer function. The reactor servo system was used to perturb the reactor by superposing a binary signal on the normal power-demand signal. Duration of the test was 15 min, and the power was perturbed by ~3% (the steady-state power was reduced from 100 to 97 MW before the ~I MW peak-to-peak power oscillations were introduced). The frequency response was determined by cross-correlating the measured power fluctuations with the reactivity input (control-rod motion). The absolute reactor power-to-reactivity gain was calculated by multiplying the measured rod motion by the differential rod reactivity worth. The gain thus obtained could be compared directly with the theoretical gain without the need for any arbitrary normalization. The experimental results indicated that the analog model was a good approximation of the reactor kinetics, including reactivity-feedback effects.

STUDIES OF POWER-TO-REACTIVITY TRANSFER FUNCTION

Since the HFIR has both negative and positive temperature coefficients of reactivity, the relative importance of each on the transfer function was determined by comparing the transfer function obtained with only negative feedback to that obtained for positive plus negative feedback. Figure I shows that the positive temperature coefficients (target water, target annulus, and control-rod water) have negligible effect on gain (2% difference at 0.18 Hz) and phase at frequencies greater than 0.1 Hz. This is due to the low percentage of total power generated in these regions (1% in target. 0.3% in target annulus, and 1% in control region). These regions are very loosely coupled to power through their respective temperature feedback reactivity coefficients, and a conclusive study of the positive feedback in the HFIR using the power-to-reactivity transfer function cannot be made. However, an analysis of the negative feedback of the fuel and core coolant water can be made.

Having determined that the predominant feedback was negative, the sensitivity of the transfer function to changes in the two negative-feedback regions was observed when the magnitude of either the fuel or core coolant temperature coefficients was doubled. Figure 2 shows the effect of doubling the fuel and water temperature coefficients to -2×10^{-5} and -1.94×10^{-4} ($\Delta k/k$)/°F, respectively. The gain function is most sensitive to changes in the fuel and water coefficients at frequencies between 0.1 and 5 Hz; the phase is more sensitive between 1 and 5 Hz.

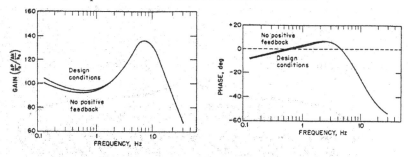

Fig. 1 Effect of positive reactivity feedback.

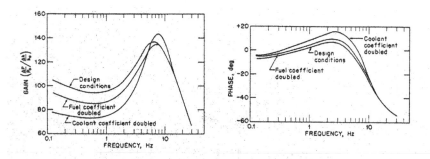

Fig. 2 Effects of coefficients of fuel and core coolant.

Note that the gain increases in the range 6 to 15 Hz when the water coefficient is doubled. Normally the gain should decrease when the negative feedback is increased. However. the phase difference or lag in the feedback can have the effect of reinforcing the reactivity disturbance (increase the gain) in some cases. such as in the HFIR. Because of the different effects caused by change in the coefficients. it may be possible to diagnose which coefficient has changed in a reactor by monitoring the gain or phase functions. The transfer-function sensitivity to changes in core coolant flow was studied to determine the feasibility of detecting flow restrictions or blockage by monitoring the transfer function. Figure 3 shows the change in the gain and phase when the core flow was reduced to one-half design flow. Reducing the flow decreases the gain for frequencies up to ~2.5 Hz. increases the gain from 2.5 to 10 Hz, and shifts the frequency at which the maximum gain occurs from 6.5 to 5.5 Hz. The phase for reduced flow is more positive in the region 0.1 to 5 Hz. Therefore it can be concluded that, for measurements of the effects of core coolant-flow changes on the reactivity feedback, the sensitive frequency ranee is 0.1 to 10 Hz for the gain and 0.1 to 5 Hz for the phase. Note that the phase seems to be a much more sensitive function of flow since the maximum gain change was ~42% at 0.12 Hz. whereas the phase changed by ~245% at 1.7 Hz.

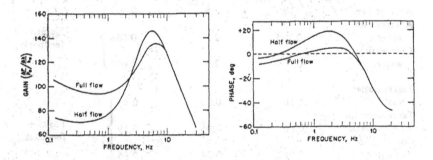

Fig. 3 Effect of core coolant flow.

STUDIES OF TEMPERATURE-TO-POWER TRANSFER FUNCTIONS

Temperature-to-power transfer functions were determined for normal reactor conditions to determine the midband gain (in °F/MW) and the break frequency f_B if f_B is known and the transfer function can be approximated by a first-order lag. the tine constant associated with temperature reactivity feedback can be determined from

$$TC = 1/ 2\pi f_B \qquad (2)$$

where *TC* is the time required for the output of a first-order lag device to reach 63% of its final value for a step change in input.

Table I summarizes the results of the temperature to-power transfer functions of the analog model. All the outlet time constants tend to be slightly longer than the actual reactivity feedback (note the difference between the time constants of average core water and outlet core water). Because of the low midband gains of the target and control regions, there is little chance to measure these transfer functions experimentally because the measurement would require too much power fluctuation to detect a temperature change above background temperature and electrical noise.

Table 1 Temperature-to-Power Transfer Functions

Temperature variable	Midband gain, °F/MW	Break frequency, Hz	Time constant,* sec	
Average fuel	1.18	2.8	0.057 ⎫	
Average core water	0.345	2.7	0.059 ⎬	Negative feedback
Core outlet water	0.690	2.5	0.064 ⎭	
Target outlet water	0.16	0.37	0.43 ⎫	
Target-annulus outlet water	0.10	0.85	0.19 ⎬	Positive feedback
Control-rod outlet water	0.082	0.24	0.66 ⎭	

*Core time constants in the Fermi reactor and 1000-MW(e) oxide breeder reactor are 0.25 and 2.4 sec, respectively (p. 195 of Ref. 4).

The results of this analog study show that the time constants for positive feedback are all longer than those for negative feedback, a fact which means that for a power excursion the negative feedback is faster than the positive feedback. This, coupled with the small amplitude of the positive coefficient, assures that the HFIR is stable.

Another result is that the time constants of the HFIR core are shorter than those of the Fermi reactor by a factor of 4 and are 40 times longer than those of a 1000-MW(e) oxide breeder reactor because the HFIR has a smaller heat capacity and a rapid coolant flow. This result is significant because it indicates that feedback information can be obtained in fast reactors without making measurements beyond about 0.075 Hz. a frequency that can probably be obtained using a technique similar to that used in the HFIR experiments.

The sensitivity of the core outlet water temperature-to-power transfer function to changes in coolant flow was determined by reducing the flow to one-half the design value. The design-flow and half-flow transfer functions are compared in Fig.4. Because the heat transfer varies as $(\text{flow})^{0.8}$, the gain increased from 0.365 to 0.53 and the break frequency decreased by $\sim(2)^{0.8}$, or 1.7. The same held true for the target region (see Fig. 5). These results, coupled with the results of the core flow reduction, indicate that by monitoring the CPSD of outlet temperature vs. power, it may be possible to detect partial flow blockages in coolant channels in reactor core sub assemblies using these measurement techniques.

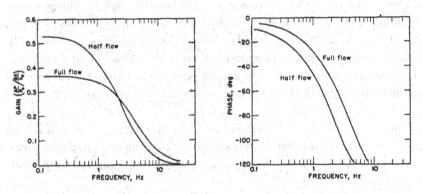

Fig. 4 Effect of blockage on core coolant temperature-to-power transfer function.

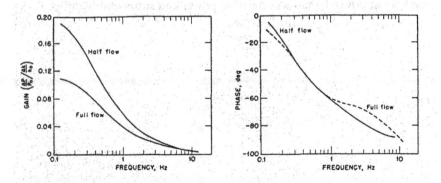

Fig. 5 Effect of blockage on target coolant temperature-to-power transfer function.

CONCLUSIONS

The following conclusions have been drawn from this study:

1. Because of the different effects caused by changes in the temperature coefficients of both the fuel and the core coolant, it may be possible to diagnose which coefficient has changed by monitoring the gain and phase functions.

2. The HFIR core thermal response is much faster than in most other reactors, including liquid-metal fast reactors, and therefore it should be much easier to apply the methods to other reactors because of the lower oscillation frequencies that must be obtained from the servo system.

3. Thermal-response time constants can be inferred by perturbing reactor power and computing the CPSD of power and temperature signals. The practical limitations of this technique are: (1) thermocouples must be installed in the correct location and have response faster than the thermal time constants, (2) induced temperature fluctuations must be greater than background electrical and temperature noise, and (3) the reactor power perturbations must contain frequencies corresponding to the time constants being measured.

4. Dynamics tests can be performed using the same instruments and equipment used for normal reactor control, i.e., using reactor power servo and control rods for reactivity perturbation instead of specially installed in-core oscillators.

5. The feasibility of detecting partially blocked flow in reactor sub assemblies at power by cross-correlating power and subassembly outlet temperature should be studied further.

REFERENCES

1. H. H. Hummel and D. Okrent. *Reactivity Coefficients in Large Fast Power Reactors,* pp. 186-188, American Nuclear Society, HinsWe, 111.,1970.

2. Ibid., p. 189.

3. R. L. Randall and P.]. Pekrul, Applications of Analog Time and Frequency Correlation Computers to Reactor-System Analysis, in *Neutron Noise, Waves, and Pulse Propagation,* Gainesville, Fla., Feb. 14, 1966, Robert E. Uhrig (Coordinator), AEC Symposium Series, No. 9 (CONF-660206), p. 359, 1967.

4. V. Rajagopal, Reactor-Noise Measurements on Saxton Reactor, in *Noise Analysis in Nuclear Systems,* Gainesville, Fla., Nov. 4, 1963, Robert E. Uhrig (Coordinator), AEC Symposium Series, No. 4 (TID-7679). p. 438, 1963.

5. F. T. Binford and E. N. Cramer (Eds.), The High-Flux Isotope Reactor, Selected Construction Drawings, USAECReport ORNL-3572, Vol. 2, Oak Ridge National Laboratory, May 1964.

6. O. W. Burke, Oak Ridge National Laboratory, personal communication.

7. D. N. Fry, On-Line Calibration of HFIR Control Rods Using the Rod Oscillation Technique, USAEC Report ORNL TM1961, Oak Ridge National Laboratory, September 1967.

8. R. C. Kryter, Application of the Fast Fourier Transform Algorithm to On-Line Reactor Diagnosis, IEEE (*lnst Elec. Electron. Eng.*), *Trans Nucl. Sci.,* NS-16(1): 210 - 217 (February 1969).

9. C. H. Chen, T. W, Kerlin, and D. N. Fry, Experience with Binary Periodic Signals for Dynamic Testing at the HFIR, *Trans. Amer. Nucl. Soc.,* 14(1) 197-198 (June 1971)..

Chapter 18

The I&C Analog/Hybrid Computer Facility

R. S. Stone

The analog computer has gone much the way of the slide rule; digital computers and calculators have displaced both of these once valuable tools. For about thirty years, however, analog machines played a vital role in the design of reactors and other dynamic systems at ORNL, and the I & C Division is where this technology came into being and was applied.

The principle of analog simulation is simple: if one wishes to know in advance the dynamic behavior of a proposed system, one builds a model and observes how it behaves. An electronic simulator puts together circuit elements whose dynamic behavior follows the same equations that describe the system under study, be it nuclear, thermal, mechanical, or biological. As ORNL scientists and engineers began the construction of a variety of experimental reactors, the I & C engineers responsible for the controls needed a way to evaluate the suitability of proposed control systems for the dynamic challenges they would face. The idea of reactor simulation by electronic circuits had appeared attractive for some time. In 1950 P.R. Bell and H.A. Straus of the ORNL Physics Division designed a machine that simulated reactor dynamics (1). This demonstrated the ability of an electronic assembly to follow the equations of a fissioning assembly. What now was needed was a programmable machine that could allow for variations in reactor design, including auxiliary cooling and control systems.

Practical and programmable analog computers first became possible upon RCA's development of a stable D.C. amplifier of high gain, the so-called "operational amplifier" (2). Using an adaptation of this circuit designed by Karl West, the Physics Division Instrument Group (later transferred to the new I & C Division) constructed 20 such amplifiers for the first Reactor Controls Analog Facility (RCAF) (3). At this time E.R. Mann led the work, aided by J.J. Stone and F. Green. Although this was a small facility by later standards, the staff achieved maximum versatility through the use of complex input and feedback circuits. Where the ultimately available commercial analog machines were limited to adding and integrating amplifiers, the RCAF was supplied with a complex RC network that allowed one amplifier to simulate the behavior of a nuclear reactor, including six delayed neutron groups. This machine provided a powerful tool for evaluating control system designs, though it is evident that each new system evaluated required programming its simulation with a soldering iron. These custom networks were built into small metal boxes attachable to an amplifier by an octal plug, thus making them quickly interchangeable.

The type of system most easily simulated by the electronic analog is one described by a series of linear differential equations with constant coefficients. In simulating a reactor one is immediately faced with a nonlinear kinetic equation wherein the effective multiplication factor must be multiplied by the power level (both being variables). To accommodate this nonlinearity Dr. E.R. Mann and J.J. Stone designed an electronic multiplier that was then constructed in the I&C shops. This design used one variable voltage to trigger gates that varied the admittance to an amplifier; the multiplying variable served as the input. In use this device required the use of two of the available amplifiers.

One of the first versions of the RCAF is shown in the picture ORNL 10147, Figure 1.

ORNL10147
Fig. 1 Early version of the RCAF as it was used in the early 50's

In the mid '50s the newly created Instrumentation and Controls Division had taken over the analog facility, and R.S. Stone joined the staff. As commercial equipment became available, I&C added an Electronic Associates 16-31R analog computer to the RCAF. This had a complement of 20 amplifiers, thus doubling the size of the facility. Also added was a nonlinear console from the Reeves Instrument Corp., comprising five electronic multipliers and four variable function generators, greatly expanding the number of nonlinear equations that could be addressed. Equally important to this expansion was the presence of a patch bay assembly with removable patch panels. Each panel is essentially like a 1600 pole single throw switch. One uses flexible patch cords to connect the elements required for a given simulation. When through with the program for the time being, this panel can be removed and stored for later use while someone else puts another panel on to run a different program. No more soldering irons!

Most reactor problems involve coolant flow. Heat transfer takes place in the reactor and in the steam generator (or other power sink), with time delays between these elements determined by the rate of coolant flow. Reactors have temperature coefficients of reactivity, making it important to have an accurate depiction of the time dependent temperature entering the core. The traditional analog approach to a time delay has been an RC time lag based upon Taylor's expansion of the function $F(t-T)$. This sounds good, but unless a facility has enough amplifiers to express a substantial number of terms in the infinite series (not usually the case), this approach gives seriously incorrect results for any but slowly developing transients. Where a change in variable is rapid (as it is in the most critical cases), the output does not show that rapid change after an appropriate delay, but instead provides a gradual change beginning immediately. In response to the need for an accurate transport delay, R.A. Dandl invented a circuit for sampling an input signal into a series of capacitors sequentially accessed by a motor driven contact, with the output read out by another contact following along on the same drive. H.J. Stripling and T. F. Sliski of I&C's mechanical group provided the mechanical design for the motor driven access to 20 circularly mounted capacitors (4). (A couple of years later we saw the exact circuit in a Russian journal, claiming a significant Soviet invention.) A picture of the Reactor Controls Analog Computing Facility is shown in Figure 2, as it existed in the mid 1950's.

Fig 2. Picture of the RCAF as it existed in the mid 50's. It was located in Bldg 4500, corridor H, until the new addition to Bldg 3500 was constructed.

At this time our facility represented an enormously powerful engineering tool. Starting with the early homemade amplifiers and adding commercial equipment over the next two decades, the department successfully designed control systems to meet the kinetic requirements of a long line of reactors. This systematic treatment began with the MTR and the ANP (nuclear aircraft) project, and was subsequently followed with work on the LITR, BSR, HRT, ORR, TSR, HPRR, MSRE, and HFIR. Much simulation went into kinetic analyses of the S.S. Savannah, EGCR, HWOCR, and PBF. Later work involved the MSBR, with safety studies on the HTGR and certain aspects of the LMFBR. (5) Prior to acquisition of our own analog/digital hybrid machine, Reactor Controls developed a detailed hybrid model of a once-through steam generator, working with the hybrid computer laboratory at the University of Illinois (Chicago). This model was used in the evaluation of control schemes for the MSBR, and is readily adaptable to other power reactors. In all, the work done represents some 400 man-years of experience in the analysis and design of reactor control systems. These systems have been successfully designed and optimized for the reactors cited, with ultimate performance well tailored to design objectives. Other studies run on the facility by us and by guest investigators have included desalination plant control, containment evaluations, fission product distributions, noise analysis (reactor perturbations), radioactive decay schemes, and biological and ecological models. During this period a number of talented staff members joined the Analysis Group. These included O.W. Burke, F.H. Clark, O.L. Smith, R. Booth, S.J. Ball, N.H. Clapp, Jr., and consultants whose expertise was sought for particular applications.

As analog technology made enormous strides in capability and applications, so too did the digital side of problem solving. By the late 60's larger and faster digital machines were widely used for a variety of applications, with modern programming methods opening their use to other than computer specialists. Indeed, the Analysis Group of Reactor Controls occasionally used digital programs when faced with needs for greater numerical accuracy or greater dynamic range than the analog was capable of supplying. The experience with the University of Illinois hybrid computer had been highly productive, and the incentives were strong to create such a system at I&C. At this stage, stand-alone digital machines were incapable of doing the on-line simulations of large systems that were routine on the RCAC. (By this time that facility had grown much larger by the addition of an Electronic Associates 221-R analog computer with 54 operational amplifiers and a lot of nonlinear equipment, plus another 16-31R like the one we originally bought. The latter two machines were surplus from other government installations.) Despite the advantage of the analog, we could not ignore the attractive features a digital computer could add with its memory, accuracy, and ability to run programs unattended.

In consequence of these considerations, in 1970 an analog/digital hybrid computer was added to the Reactor Controls computer lab. This comprised an Applied Dynamics analog computer with 153 operational amplifiers, 16 electronic multipliers, and 128 digitally adjustable coefficient devices. There were 20 digital to analog and 24 analog to digital information channels, plus sense and interrupt lines between the two machines. The three older analog consoles still constituted a powerful computing facility, but were of limited usefulness in the hybrid mode because of the low-speed relay switching they used, and the relatively narrow bandwidth of the amplifiers. The digital computer was a Digital Equipment Corp. PDP-10, a medium size machine for its time. It had 32K of 36-bit words of 1.8-microsecond memory, and included hardwired floating point and multiply/divide. This was rather a pitiful computer by today's standards, but had today's digital computers been available, we wouldn't have needed the hybrid.

Once installed, the computer was usually run in the hybrid mode. The systems studied went on the analog as usual, but the digital adjunct offered so many advantages that it was virtually always tied in. It quickly set the coefficient potentiometers, for one thing, a tedious and time-consuming task when done by hand. It also provided wonderful transport lags. Just store a temperature in the digital machine and tell it when you want it back. Complex functions could be generated; the steam tables could be stored to read out dependent variables on demand. Or the problem could be run on the digital machine, using the analog as a differential equation solver.

Soon after the hybrid machine became available, a huge analysis program was initiated under contract with the NRC. This was an in-depth study of two commercial nuclear power plants, one Babcock and Wilcox, the other Combustion Engineering. The object was to determine the safety implications of control systems in these plants.

Fig. 3 The Reactor Controls Hybrid/Analog Computing Facility. It was housed in Bldg 3500, on the 2nd floor.

This was a major effort that occupied four engineers full or part time for several years. (6) (7) It was not a study that would have pleased E.P. Epler, had it been done when he still headed Reactor Controls. Ep's strong opinion was that the safety system must be capable of handling any emergency event that a control system could initiate, even if such a system "fell off onto the floor" or took off in a direction 180 degrees from where it was supposed to go. The only thing the control system was not allowed to do was fail in a dangerous way on a regular basis, since this could statistically lead to an emergency occurring during the extremely rare condition wherein the safety system had an undetected failure. This is called the Anticipated Failure Without Scram, or ATWS, and has had programs dedicated to its prevention and mitigation. Our analysis was extremely thorough, examining every conceivable malfunction in every portion of extraordinarily complex control systems. Although we detected some problems, and uncovered some questionable operating practices, we found no horrible accident waiting to happen. Perhaps that is why the follow-on program to look at Westinghouse and General Electric plants was never funded.

In addition to this major program, the studies that had previously been done on the analog now moved to the hybrid. Nuclear desalting, artificial intelligence, noise analysis, etc., now ran on the hybrid as its advantages became evident. As part of a study of heat transfer in a nuclear fuel pin, we used a hybrid equivalent of the DOE's THETA digital program. We could in an hour make a run on the hybrid that would take the biggest digital machine all day to complete, and with less detail than the hybrid. Our machine cost about $500,000, the big digital machines of the day cost several million. I believe we saw the specter of things to come when the DOE preferred to run their calculations on the digital machines; they had more adherents rooting for them. With the rapid progress in digital hardware and software, eventually that technology was certain to make analog simulation obsolete. It seemed too bad it happened while the hybrid machine was still cheaper, faster, and gave more accurate results.

Be that as it may, for over 30 years I&C's analog and analog/digital hybrid computers gave vital engineering answers available from no other source. These machines and the problems they solved were in the best tradition of I&C's contributions to science and engineering.

References

1. P.R. Bell and H.A. Straus, "Electronic Pile Simulator", Rev. Sci. Instr. 21 (8), 760-63 (August 1950)

2. Goldberg, E.A. "Stabilization of Wide-Band Direct-Current Amplifiers for Zero and Gain", RCA Review 11, 296 (1950)

3. J.J. Stone and E.R. Mann, "Oak Ridge National Laboratory Reactor Controls Computer", ORNL-1632 (Reissued March 1954)

4. R.S. Stone and R.A. Dandl, "A Variable Function Delay for Analog Computers", IRE Transactions on Electronic Computers, 187-189, (September 1957)

5. R.S. Stone, The Controls Department Hybrid Computer Hardware and Applications, ORNL-TM-3101 (August 1970)

6. R.S. Stone, F.H. Clark, O.L. Smith, A.F. McBride, N.E. Clapp, Jr., R.E. Battle, An Assessment of the Safety Implications of Control at the Oconee 1 Nuclear Plant, NUREG/CR-4047, ORNL/TM-9444 (March 1986)

7. S.J. Ball, R.E. Battle, E.W. Hagen, L.L. Joyner, A.F. McBride, J-P.A. Renier, O.L. Smith, R.S. Stone, An Assessment of the Safety Implications of Control at the Calvert Cliffs-1 Nuclear Plant, NUREG/CR-4265, ORNL/TM-9640 (April 1986)

Chapter 19

The Analysis of Nuclear Reactors, According to their Noise Spectra

Ray Adams

I attempted to get some of the principal parties to write about this part of the I&C Division's contributions to the Reactor Controls legacy. Each has declined, so I have written what I remember of this effort that I became aware of in the late 1960's and early 1970's. My first recollection of I&C's Bob Kryter and Dwane Fry, in this area was when they applied a multi-channel analog tape recorder [Ref 1] to record noise signals that occurred in a nuclear reactor, so that they could analyze the signals off-line, for possible clues to various reactor behavior. This portable tape recorder was used to record signals from various reactors around the US, and quickly led to the acquisition of on-line computer analysis tools and the refinement of the use of random noise signals in the analysis of nuclear reactors.

Both Kryter and Fry had earlier published on the use of FFT analysis papers [Refs 2 & 3] for nuclear reactor diagnostics. Their work in this area led to an elite small conference which brought together experts in the field. The conferences (several of them followed) came to be named SMORN, which stood for Specialists Meeting On Reactor Noise. The first of those conferences was held in Rome, Italy, in 1974, the second in Gatlinburg, TN, in 1977, and the conferences have continued at intervals of 3 to 5 years. The most recent SMORN Conference I have found was held in 2002 and it appears to have morphed into a conference on Reactor Surveillance and Diagnostics.

Reactor Noise analysis turned out to be useful for detecting other things inside an operating reactor. For example what if, during construction, a worker had left a wrench or other tool, inside the containment, or what about effects due to wear of internal reactor structures? That wrench could cause untold havoc, banging around with the cooling water, or if parts designed to hold structures steady did not. This is not so far fetched as it may seem and in 1977, Kryter (and others) published a paper that focused on loose part monitoring [Ref 4] by various means. An earlier paper [Ref 5] by Fry, Kryter (and others) in 1975 focused on the detection of vibration of inner parts of a nuclear power reactor core.

Other papers were published in the area of Noise diagnostics of Power Reactors by Fry and Kryter (with others) as well as a NUREG report [Ref 6].

I consider Dwane Fry and Bob Kryter to have been the leaders of the I&C contributions of the use of random fluctuations of Neutron Noise to the diagnostics of Nuclear Reactor problems. I knew both Bob Kryter and Dwane Fry as solid contributors to this field and they were perhaps among the earliest contributors world-wide, as well.

Dwane Fry became the head of the Reactor Controls Department in the I&C division, whereas Bob Kryter preferred to remain a researcher, out of the management chain.

ORNL101321
Dwane Fry as a researcher.

ORNL05458-78
Bob Kryter

Bob Kryter was always a researcher. He thoroughly studied the principles of the analysis of random signals and applied them to nuclear reactor processes.

REFERENCES

1. D. N. Fry and R. C. Kryter. Paragraph 8.22 of 1974 I&C Annual Progress Report ORNL 5032

2. R. C. Kryter, Application of Fast Fourrier Transform Algorithm to On-Line Reactor Diagnosis IEEE Trans. Nucl. Sci.; NS-16: No 1, 210-17 (Feb 1969) .; From 15. nuclear science symposium; Montreal, Canada (23 Oct 1968).

3. Fry, D. N.; Kryter, R. C.; Robinson, J.C. On-Site Diagnostics at Pallisades Nuclear Power Station; (Oak Ridge National Lab., Tenn. (USA)), 1973

4..Kryter, R. C.; Ricker, C. W.; Jones, J. E. (Oak Ridge National Lab.,TN) Loose-Parts Monitoring: present Status of the Technology, etc. (ERA citation 03:010095) Sponsor: ERDA, 1977, 6p

5. Fry, D. N.; Robinson, J. C.; Kryter R. C.; Cole, O. C. (Oak Ridge National Lab.,TN) Core Component Vibration Monitoring In Bwrs Using Neutron Noise, 1975, 5p

6. Fry, D. N.;Kryter, R. C.; Robinson, J. C. (Oak Ridge National Laboratory) Noise Diagnosis for Safety Assessment, Standards, and Regulation ORNL/NUREG/TM-278, Nov 1976, 31 p.

Introduction To Section 4

Ray Adams

Contained in this section are a few papers that have been chosen from those originally selected for inclusion in a volume of the ORNL Review special issue on Measurements and Controls. Approximately 40 articles were selected for that issue of the ORNL Review. But that issue will not be published due to an unfortunate loss of funds at a time when the papers were nearly ready for print. The editors of this book have chosen to publish a few of those papers that seemed to us to capture some of the flavor of the work in the I&C Division of the late 1990's, and that represent the 1990's contributions of the I&C Division as Beyond the Edge of Technology. Thanks to Carolyn Krause (editor of the ORNL Review) and to the authors of the papers that are published in this section, for their permission and cooperation in publishing the papers here.

Papers published in this section:

New Instrument Expected To Improve Productivity in Semiconductor Industry by Carl Remenyik

Identifying Nuclear Weapons Material and Confirming Its Conversion to Reactor Fuel by John T. Mihalczo and Timothy E. Valentine

Oak Ridge's Heartbeat Detector: You Can Run But You Can't Hide by Stephen W. Kercel

Adaptation of the Fuzzy K-Nearest Neighbor Classifier for Manufacturing Automation by Kenneth W. Tobin, Shaun S. Gleason, and Thomas P. Karnowski. Note that Ken Tobin is the Director of the reincarnation of the I&C Division, The new Measurement Science and Systems Engineering Division.

Particle-Physics Electronics by Charles L. Britton and Alan L. Wintenberg

Chapter 20

New Instrument Expected To Improve Productivity in Semiconductor Industry

Carl Remenyik

ABSTRACT: ORNL has developed a calibrator to improve the accuracy of measurements of the flow of gases essential to the production of integrated semiconductor circuits for computers and other electronic devices. A modified version of the device could also reduce worker exposure to toxic and corrosive gases used in the semiconductor industry.

ORNL has developed an instrument primarily for the calibration of gas flow controllers used in the semiconductor industry. This instrument collects and weighs the gas that flowed through the controller to be calibrated. The sensitive balance weighing the gas is relieved of the weight of the heavy cylinder collecting the gas by submerging it in water and offsetting its weight with the cylinder's buoyancy. This instrument received an R&D 100 award in 1995 from *R&D* magazine.

Processes for manufacturing semiconductor devices used in televisions, radios, computers, and many other electronic devices require accurately controlled flows of many different gases into a vacuum chamber. Some of the gases used in industry are among the most corrosive and toxic, such as tungsten hexafluoride, silane, hydrogen bromide, chlorine, dichlorosilane, and boron trichloride. A plate or "wafer" inside the vacuum chamber serves as a substrate on which the semiconductor devices are formed. By various means, the molecules in the gases are caused to deposit in layers at selected locations on the surface of this substrate; to diffuse into existing layers to change their electric, thermal, or chemical properties; or to etch off some material from the surface. The final product is an integrated semiconductor circuit.

As the speed and data-storing capacity of computers has increased over the past few decades, the required dimensions of circuit components are made ever smaller and the components are packed more densely on the wafers. At the same time, the wafers have been made bigger. These developments have increased the cost of producing finished wafers so much that their dollar values are now typically six-digit numbers for each wafer. It is not surprising, then, that manufacturers are very concerned about the rates at which rejects turn up among their products. To help reduce the reject rate related to improper gas flow, ORNL has developed a calibrator to improve gas flow measurements.

Calibrating Gas Flow Controllers

The quality of semiconductor circuits depends greatly on the accuracy with which the flow of gases into the vacuum chamber is controlled. The most frequently used instruments that regulate gas flows are mass flow controllers (MFCs). They measure and control the mass of the gas that flows through them. MFCs must operate with an error of not more than 1% to make the desired reproducibility and quality of the product possible. Before MFCs can be used in fabrication systems, they must be calibrated. The calibration must be even more accurate than the device it calibrates. Experience teaches us that the calibration standard should not have an error greater than about 0.1% if a device needs to be calibrated with an error of 1%.

Manufacturers and laboratories use various calibration methods. Several methods determine the rate of mass flow indirectly. With some, the gas pressure, temperature, and volume must be measured, and, in addition, the gas constant and compressibility factor must be known. From these, the mass is calculated with a thermodynamic relationship. These methods suffer from the fact that each of the quantities entering the calculations contain some measurement error, affecting the results cumulatively. Furthermore, some of these methods cannot be applied to toxic or corrosive gases. Other indirect methods of calibration themselves require calibration by one of the other methods and, therefore, cannot be more accurate.

Some methods of calibration weigh gases leaving or entering the calibrated device during a measured time. Measuring weight may be regarded as equivalent to measuring mass and, therefore, such gravimetric methods of calibration may be considered direct methods. In a typical procedure, a pressure cylinder is evacuated, weighed, and then connected through pipes and valves to the calibration setup with the device to be calibrated. During a measurement, the elapsed time is measured from the instant the valve is opened, when gas begins to flow into the cylinder, until the valve is shut. Then, the cylinder is disconnected and weighed again. The increase in weight is the weight of the collected gas. From the measured weight and time, the average mass flow rate of the gas streaming through the device and into the cylinder can be calculated. The measurement can be performed with an initially pressurized cylinder from which gas is discharged during the measuring process.

Generally, the error of a precision balance is a percentage of the full range of the balance. When the pressure cylinder is weighed before and after the operation, the weight of the gas—the difference between the two weighings—is determined with an error corresponding to the total weight on the balance, which includes the heavy steel cylinder. If the amount of the collected gas weighs only a small fraction of the weight of the cylinder (e.g., a few tenths of an ounce), this error is too large a percentage of the gas weight. Because of this problem, the measurement procedure must continue until a sufficient amount of gas has been collected. At low flow rates, the collection may require many hours or even days. During all that time, the entire system has to operate steadily with very small drifts and oscillations of temperature, pressure, and instrument settings because, at the end, only the average flow rate over the entire time can be calculated, and therefore the accuracy of the calibration depends on the steadiness of the operating conditions. One such measurement determines just one point on the calibration curve.

This method is time-consuming and cumbersome because the cylinder must be disconnected from the pipes of the calibration system for every weighing. This method introduces occupational hazards when the gas is toxic or corrosive, necessitating purging of all or part of the system before the pipe connection may be opened. The length of the required time makes calibration very costly, even for one gas. But mass flow controllers respond differently to different gases, necessitating separate calibration for each gas (which is prohibitively expensive). Some existing methods allow calculation of approximate calibration coefficients from the results for other gases, but these methods are not satisfactory.

ORNL's Contribution

The semiconductor industry needs improved calibration technology to overcome the shortcomings of existing methods, to avoid worker exposure to toxic gases; to make measurements involving highly reactive gases; and to improve the accuracy and speed of calibration, thus increasing the safety and efficiency of the industry's manufacturing operations. This need prompted an effort in ORNL's Instrumentation and Controls Division to develop a new instrument that:

· measures the weight of the process gas directly,

· eliminates the need for disconnecting the instrument from the flow system during the calibration process,

· reduces the time needed for a measurement by orders of magnitude,

· may be used with the corrosive and toxic process gases involved in the fabrication of semiconductors, and

· increases accuracy tenfold over the accuracies of commonly used instruments.

The new ORNL-developed gravimetric calibrator meets these criteria. The first prototype was made of stainless steel, thus allowing the instrument to resist attack by reactive gases that cannot be handled by other types of calibrators. Our instrument is being used in ORNL's Mass Flow Controller Testing Laboratory, which was originally set up to carry out work for SEMATECH, a research consortium of the semiconductor industry, and is now a user facility available for industrial firms, universities, and other organizations. The patented instrument has been in operation since 1994, and it is the only instrument of its type known to exist.

The idea behind this invention is to make an empty steel vessel almost weightless by submerging it in water to balance its weight by buoyancy. In other words, the water pushes against the 23 kilogram (50-pound) container enough so that it feels no heavier than one-tenth of an ounce. The submerged container is suspended from a load cell balance, an electrical device that can measure weight by sensing the elastic deformation of one of its components caused by the weight. It works like a strain gage, whose electrical resistance is increased the more it is stretched or bent (thus decreasing the measured voltages passing through it by an amount related to increases in applied mechanical force). The sensitive and delicate load cell weighs only 3 to 4 grams (about the weight of a sheet of typewriter paper). The one used here has a load-bearing limit of 50 grams and would instantly collapse under the container's weight. Therefore, the container is submerged in water and its weight is balanced by buoyancy. Tare weights ensure that there is a small load, 3 to 4 grams, on the load cell even when the container is evacuated. Without these small weights, buoyancy might lift the container off the load cell.

ORNL3823-95

Fig 1 Author Carl Remenyik checks a mass flow controller that feeds gas to the gravimetric gas flow calibrator that he developed for the semiconductor industry. Below is the submerged cylindrical vessel in which a corrosive or toxic gas can be safely weighed. Next to his left shoulder is the load cell balance from which the submerged vessel is suspended. For this device Remenyik received an R&D 100 award in 1995. Photograph by Bill Norris.

The main part of the gravimetric calibrator (shown in Fig. 1) is a closed, double-walled cylindrical container that is nearly 2 meters (6 feet) long. The space enclosed by the inner wall communicates with the outside through a capillary tube. This tube connects with a system of gauges, flow controllers, and remotely controlled valves, which can ultimately connect the container either to the instrument to be tested or to a vacuum pump. Or the valve can be closed, isolating the system. Figure 2 is a schematic which shows the various elements more clearly.

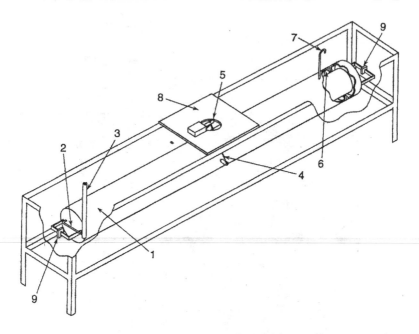

Fig 2 Schematic of the gravimetric gas flow calibrator showing; 1) the container, 2) the connecting capillary tube, 3) the vertical pipe connection, 4) the support frame, 5) the load cell, 6) the support ring for container, 7) the capillary vent pipe, 8) the support plate for the load cell, 9) the tare weight.

In preparation for a series of measurements, the container is first evacuated. The measurement of the gas flow rate through the device to be calibrated begins when a valve switches the flow from a by-pass into the container. The load cell continuously measures the increasing weight of the accumulating gas and sends electrical signals to a computer, which records the increasing weight 2000 times a second. The measurement may last seconds or minutes, depending on the magnitude of the flow rate, and it is terminated by switching the valve back to its initial bypass direction. The next measurement may follow immediately.

As the accumulating gas applies an increasing weight on the load cell, the cell deflects slightly, lowering the container. Even though the maximum displacement of the container is only a fraction of the thickness of writing paper, it still twists the capillary tube, resulting in a small change of the force acting on the load cell. That is an error, but the capillary tube was designed and arranged so that the error caused is always more than 10,000 times smaller than the weight of the gas; because it is much smaller than the specified error, it is ignored. However, this small effect can either be calculated or measured, and the error caused by it in the results may be corrected, should that ever become necessary.

The double walls of the container also affect the accuracy. When the container is being evacuated, its dimensions shrink a little even though it is made of steel. The force of buoyancy acting on an object in water changes if the volume the object occupies changes. When gas enters the container during a measurement, the pressure rises inside the container and expands it, increasing its buoyancy and causing a false change in the gas weight. This source of error is eliminated by inserting the container inside a shell that is large enough to leave a gap around the container. When this assembly is submerged in water, the buoyancy is determined by the outside dimensions of the shell, which remain unaltered by changes in the volume of the container inside.

A second prototype of this instrument is now under development. Experience with the first instrument and changes in the semiconductor industry have prompted these improvements. The second prototype is more versatile and much smaller than the first instrument, and possibly compact enough to become a portable, table-top device. If the improved instrument works as planned, it will likely be adopted for use by the semiconductor industry.

BIOGRAPHICAL SKETCH

Carl Remenyik was a consultant with ORNL's Instrumentation and Controls Division. He also has been a professor of fluid mechanics at the University of Tennessee at Knoxville. He made important contributions to the development of the cytriage (blood analysis monitor) in the 1960s, and he worked with Jim Hylton to develop the gas flow calibrator for which he received an R&D 100 award in 1995. He has a PhD. degree in engineering sciences from Johns Hopkins University. He first began working at ORNL in 1966 when he joined the Reactor Division. Before coming to Tennessee, he worked for the Glenn L. Martin Company, which eventually became Lockheed Martin Corporation.

Chapter 21

Identifying Nuclear Weapons Material and Confirming Its Conversion to Reactor Fuel

John T. Mihalczo and Timothy E. Valentine

ABSTRACT: Researchers at ORNL and the Oak Ridge Y-12 Plant have developed a non intrusive nuclear weapons identification system for confirming that nuclear fissile material (uranium and plutonium) has been removed from dismantled nuclear weapons and that the fissile material is being safely stored in designated vaults. The group has also developed a fissile mass flowmeter for verifying that weapons-grade uranium is being blended down to reactor-grade uranium.

When a nuclear weapon is dismantled, how do we know that all the fissile material has been removed? How can we be sure that the removed uranium-235 (^{235}U) or plutonium-239 is present in the designated storage area? How can it be verified that a nation has no more nuclear weapons or nuclear material than it declares and that it is complying with bilateral treaties? How can a facility be sure that a shipment contains the nuclear material ordered or that nuclear material is absent from a container? How can we be certain that highly enriched uranium is in a storage vault or that a train car load of spent nuclear fuel won't go critical, causing an inadvertent release of radiation?

To answer these questions reliably, a group in ORNL's Instrumentation and Controls (I&C) Division along with other researchers at ORNL and at the Oak Ridge Y-12 Plant (see sidebar for complete list of names) have been developing a nuclear weapons identification system (NWIS) since 1984. This system is currently in use at the Y-12 Plant. NWIS can be used to verify that nuclear weapons have been dismantled and that their nuclear contents have been properly stored. Under the latest proposed Strategic Arms Reduction Treaty (START III), dismantlement of nuclear weapons and storage of

271

nuclear materials removed from weapons can be monitored for the first time. NWIS has been demonstrated to Russian scientists using nuclear weapons components at the Y-12 Plant, and the technology is being transferred to them so that they can use it to verify that fissile material has been removed from their dismantled nuclear weapons. In addition, NWIS has also been used locally to verify that certain containers returned to the Y-12 Plant from the military had no highly enriched uranium.

NWIS uses neutrons from the spontaneous fission of low-intensity californium-252 (^{252}Cf) sources to "interrogate" the weapons component in question. A ^{252}Cf source in an ionization chamber provides a timed source of fission neutrons that enter the weapon or component, inducing fission internally in any fissile material present. Two detectors on the opposite side of the weapon detect the emitted gamma rays and neutrons. These radiation signals consist mostly of directly transmitted and fission-induced gamma rays, as well as directly transmitted, scattered, and fission-induced neutrons. Data from the source ionization chamber and the two detectors are processed to determine the time required for the various types of neutrons and gamma rays to reach the detector channels and the frequency components of the signals. The data analysis produces a fingerprint, or signature, which can be compared to reference fingerprints. This very robust signature consists of 19 functions of frequency and time that indicate whether the component contains nuclear material.

Improving NWIS

The first attempts to characterize nuclear fissile materials involved the use of pulsed neutrons from an accelerator, pioneered in the 1940s at Los Alamos National Laboratory. A similar system was used at the Y-12 Plant, which manufactured parts for nuclear weapons and stored weapons material. In 1968 ORNL and Y-12 researchers developed a replacement for the accelerator that worked just as well as the accelerator system and had the advantages of being somewhat more portable, more reliable, and easier to operate. However, until 1996, the extensive use of this NWIS methodology was limited because the hardware was bulky and cumbersome, its sampling and processing rates were low, and its measurement times were long.

In 1996, an ORNL/Y-12 team developed an improved NWIS processor based on a high-speed gallium arsenide chip that does not suffer the deficiencies of the earlier processors. The system's size and measurement times have decreased and its sampling and processing rates have been increased by 1000 times over that available just 10 years earlier. For some systems, measurements can be completed in times as short as 10 seconds. In late 1996, the development of a laptop-based version of the processor was initiated (see Figs. 1 and 2).

ORNL129601-97

Fig. 1. Eric Breeding, Jim Mullens, and Mike Paulis show the NWIS package and Dolch laptop computer used to determine uranium and plutonium concentrations in various weapons components and storage vaults.

ORNL129602-97

Fig. 2. Eric Breeding holds the electronic board that is the heart of the more compact version of the NWIS, while he and Jim Mullens examine an NWIS detector.

In addition to improving processor hardware, automated pattern recognition algorithms for confirmatory measurements are being implemented. The team is working on packaging the detection systems to make them more portable.

The improved NWIS has several advantages. First, it has high sensitivity— small changes in fissile masses and configurations (arrangement of nuclear materials) produce large changes in signatures. Second, some signatures are not sensitive to background radiation from nearby materials or sources, such as those present at facilities like the Y-12 Plant. Thus, NWIS is useful for verifying nuclear fuel storage configurations or for tracking nuclear weapons components through the first stage of dismantlement (see sidebar), because the presence of the plutonium parts on the assembled system does not affect some signatures for the nuclear weapons component. Third, NWIS can be non intrusive—it does not reveal weapon design information, making it useful for verification of compliance with bilateral treaties or for applications by the International Atomic Energy Agency (IAEA). Finally, the system is very difficult to deceive.

The Value of NWIS

To date, NWIS has been used to perform measurements on 15 different weapons systems in a variety of configurations, both in and out of containers (see Fig. 3).

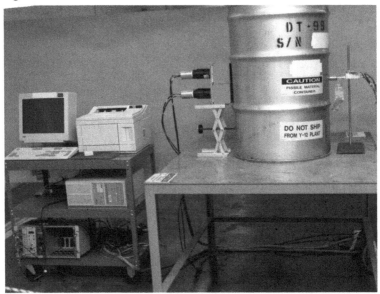

Y-12-323207

Fig. 3. This photograph shows the components of an older version of the NWIS, which is being used here to determine the concentrations of fissile materials in a storage drum. The newer version is more compact and portable.

These systems included weapon components at the Y-12 Plant, as well as plutonium parts and fully assembled weapons systems ready for deployment at the Pantex Plant in Amarillo, Texas

These measurements have shown that NWIS can identify nuclear weapons or components; nuclear weapons or components can be distinguished from mockups in which fissile material has been replaced by non fissile material; nuclear weapons components can be tracked through the first stage of dismantlement; omissions of small amounts (4%) of fissile material can be detected; and changes in internal configurations can be determined. NWIS has been used to verify that mockups of nuclear artillery shells used in military training contain depleted uranium, not enriched uranium. These mockups, or B33 components, were brought in 512 containers to the Y-12 Plant for inspection using NWIS. Verification was performed on as many as 32 containers in one 8-h shift.

Measurements have shown that NWIS can be used to identify (by type) nuclear weapons or components and fissile material retrieved from dismantled weapons. It can be used to track components and materials before and after weapons dismantlement and to help IAEA and other agencies verify declared storage, thus making it particularly useful for arms control and nonproliferation, domestic safeguards, and international agreements. Use of this system is continuing in the weapons storage program at the Y-12 Plant. In 1997 NWIS was successful in identifying plutonium metal components in tests at the Pantex Plant and in blind tests at Los Alamos National Laboratory. In these test the high sensitivity of NWIS to plutonium mass was demonstrated.

In addition to nuclear weapons applications, NWIS signatures can be used for nuclear materials control and accountability, fissile mass assay and subcriticality of spent fuel, confirmation of shipper declarations, and nuclear process monitoring. It has been used to verify quantitatively the nuclear criticality safety of highly enriched uranium in storage vaults at the Y-12 Plant

In short, this state-of-the art measurement system is also making an important contribution to both nuclear safety and national security.

SIDEBAR 1:

Megatons to Megawatts: Monitoring Blend-down of Weapons-grade Material to Fuel for Power Reactors

Under the Highly Enriched Uranium (HEU) Purchase Agreement signed February 18, 1993, by the U.S. and Russian Federation governments, our nation committed to purchase 500 metric tons of HEU extracted from dismantled Russian nuclear weapons. The United States will purchase the

HEU over 20 years for about $12 billion. The payments will be made after batches of HEU have been extracted and diluted, or downblended, to low-enriched uranium for use as power reactor fuel. This agreement has two advantages. It will prevent the uranium in these weapons from being recycled to make other nuclear weapons, thus contributing to peace. In addition, it will provide economic benefits to both countries, mainly by extending each nation's uranium supplies for energy production.

As part of the agreement, it must be confirmed that downblending has occurred as specified. An instrument developed at ORNL will be one of two used to verify that the Russians have properly blended down the purchased HEU, the first delivery of which arrived on June 23, 1995.

Blending the uranium is accomplished by first converting the removed uranium metal from weapons into uranium hexafluoride gas. The uranium metal from weapons contains about 90% uranium-235 (^{235}U). This uranium hexafluoride (UF6) gas is them mixed with UF6 enriched to 1.5% ^{235}U to give a UF6 product with an assay of 4.5% to 5% ^{235}U. The resulting Lowly Enriched Uranium (LEU) will be sent to the Portsmouth Gaseous Diffusion Plant in Ohio, which will then ship it to fuel fabricators for production of reactor fuel elements. To verify that the HEU is blended down, a special measurement system that could be deployed without cutting pipes was needed.

To meet this requirement, an ORNL team led by John Mihalczo, Jim Mullens, and Jose March-Leuba developed, tested, demonstrated, and implemented the blend-down fissile mass flowmeter (see Fig. 4).

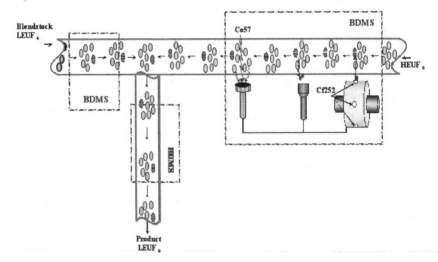

Fig. 4. Schematic of ORNL's blend-down fissile mass flowmeter and of the blend-down process

A prototype was demonstrated to DOE and the Department of State and to representatives of the Russian Ministry of Atomic Energy. The United States has negotiated the installation of this equipment in Russian facilities involved in the blend-down of HEU as a way to provide confidence that the Russian Federation is abiding by terms of the agreement.

The flowmeter contains a californium-252 (^{252}Cf) source that introduces neutrons periodically into the pipe carrying UF6 of a particular enrichment. The ^{252}Cf source is contained in a polyethylene moderator around the pipe in which the UF6 is flowing. Between the moderator and the pipe, a neutron-absorbing material is alternately inserted and removed, producing a modulated source of thermal neutrons that induces fission in the UF6 gas, producing waves of activation (delayed gamma ray emissions by uranium atoms in the gas along the pipe) that are detected downstream. The time delay between the activation and the detection downstream, and the distance between the source and detectors, allows measurement of the gas flow velocity. The amount of the signal downstream at the detector is proportional to the concentration of ^{235}U in the gas. (The more uranium, the more gamma rays in the signal.) From these two quantities, the fissile mass flow is obtained.

The blend-down fissile mass flowmeter is undergoing calibration and verification demonstrations with 1.5 wt % ^{235}U-enriched gas at the Paducah Gaseous Diffusion Plant in Kentucky. It will be installed in Russia to monitor blend-down of the U.S.-purchased uranium. After downblending at two Russian plants, the LEU product is loaded into industry-standard cylinders, transported by rail to St. Petersburg, Russia, and shipped to the Portsmouth Gaseous Diffusion Plant in Piketon, Ohio. There the LEU from Russia may be further downblended to change the enrichment level or the cylinders may be sold as received to the five U.S. nuclear fuel fabricators.

The flowmeter has been demonstrated to Russians as part of their training so that they can safely install and operate it at the Russian blend-down facilities. Jim McEvers and William H. Sides are project managers for the fabrication and installation of the flowmeter in Russia.

This program has brought international recognition to ORNL. The Laboratory received a substantial amount of funding for the development of this instrument for use in the Russian blend-down facilities. While reducing nuclear weapons-grade uranium to meet our national goal in arms control and nonproliferation, the ORNL instrument is helping the United States extend its supplies of nuclear fuel for power production, helping meet yet another federal goal.

SIDEBAR 2

NWIS Spells Relief for One Criticality Safety Concern

From 1946 to 1985, the former Oak Ridge Gaseous Diffusion Plant made a product enriched in ^{235}U for use as fuel in nuclear power plants and for nuclear weapons. Some of the uranium was deposited in pipes through which uranium hexafluoride once flowed. Recently, one deposit known to contain a large amount of material was a criticality safety concern.

The ORNL team was asked to use NWIS to locate and determine the shape and size of nuclear material deposited in the pipe and the hydrogen content, all of which is of interest to specialists concerned with nuclear criticality safety. NWIS measurements showed that the deposit was located at the top of the pipe, although previous gamma ray spectrometry measurements showed that the deposit was uniform. Based on these measurements, a hole was drilled from the bottom instead of the top to insert a video camera to view the deposit. Video images confirmed that the location and size of the deposit had been accurately determined by NWIS. The determination by NWIS that the hydrogen-to-uranium ratio was low alleviated the criticality safety concern.

This technology has a wide variety of applications other than these already described. Examples include subcriticality measurements for spent light-water reactor fuel, nuclear process monitoring and control, monitoring and control of accelerator-fission multipliers for transmitting waste from nuclear power reactors, many other applications in nuclear material control and accountability, monitoring the startup of power reactors, and in-plant measurements for nuclear criticality safety. A variation of the NWIS technology could also be used to monitor the control of actinide waste burners in which accelerator beams transmute high-level radioactive reactor waste to low-level radioactive material. It can also monitor other systems, such as accelerators that produce tritium. It can be used to characterize the fissile material removed in the cleanup of ORNL's defunct Molten Salt Reactor Experiment and to guide the safe and cost-effective storage of spent fuel at DOE's Savannah River Site. Contracts are in place to transfer this technology to two Russian nuclear weapons design laboratories—Arzamus-16 and Chelyabinsk-70.

SIDEBAR 3

Contributors to the Development of Nuclear Materials Measurement Devices

The following ORNL researchers contributed to the development of NWIS: E. D. Blakeman, J. E. Breeding, M. Brock, W. L. Bryan, R. I. Crutcher, M. D. Cutshaw, P. P. Deporter, M. S. Emery, S. S. Hughes, D. E. Hurst, U. Jagadish, R. W. Jones, J. K. Mattingly, J. A. McEvers, D. E. McMillan, J. T. Mihalczo, V. C. Miller, J. A. Mullens, V. K. Pare, M. J. Paulus, L. D. Phillips, J. A. Ramsey, S. L. Shelton, M. C. Smith, G. W. Turner, T. Uckan, T. E. Valentine, R. I. Vandermolen, and L. C. Watkins. Also working on the project were consultant R. B. Perez and University of Tennessee (UT) researchers D. L. Bentzinger, L. Clonts, B. Puckett, J. Vann, and M. S. Wyatt.

The following ORNL researchers contributed to the development of a fissile mass flowmeter for the blend-down of highly enriched uranium: A. J. Beal, M. Brock, K. N. Castleberry, R. E. Cooper, J. C. Cox, P. P. Deporter, K. W. Drescher, D. M. Duncan, D. R. England, R. K. Ferrell, A. C. Gehl, J. A. Hawk, S. J. Henley, J. S. Hicks, W. Holmes, D. E. Hurst, R. E. Hutchens, J. O. Hylton, R. W. Jones, J. A. Kilby, W. W. Koch, R. Lenarduzzi, R. D. Lively, B. R. Maggard, J. A. March-Leuba, C. W. Martin, J. K. Mattingly, J. A. McEvers, D. E. McMillan, J. T. Mihalczo,

J. A. Mullens, J. K. Munro, M. J. Paulus, S. L. Shelton, C. L. Sowder, J. D. Stooksbury, J. C. Turner, T. Uckan, T. E. Valentine, R. A. Vines, K. S. Weaver, and S. E. Worley. Contributors from the East Tennessee Technology Park (formerly the Oak Ridge K-25 Site) were D. H. Powell, S. A. Sampsel, and J. N. Sumner. Another contributor was D. L. Bentzinger of UT.

BIOGRAPHICAL SKETCH

John T. Mihalczo, a Lockheed Martin Corporate Fellow in ORNL's Instrumentation and Controls Division, is one of the developers of the nuclear weapons identification system (NWIS). He invented the [252]Cf-source-driven noise analysis method which has been applied internationally in nuclear criticality safety and nuclear safeguards applications. He has 40 years of experience with measurements of fissile materials. Mihalczo has a PhD. degree in nuclear engineering from the University of Tennessee at Knoxville, where he is a professor of nuclear engineering.

Timothy E. Valentine is a research staff member in ORNL's Instrumentation and Controls Division. He received his B.S., M.S., and PhD. degrees from the University of Tennessee at Knoxville. He came to ORNL in 1990. His main research areas are in nuclear safeguards, nuclear criticality safety, and neutron transport theory. He is one of the developers of the nuclear weapons identification systems and has performed a variety of safeguards and subcriticality measurements.

Chapter 22

Oak Ridge's Heartbeat Detector:
You Can Run But You Can't Hide

Editor's Note: The Paper by Stephen Kercel has been augmented here, by material that was submitted when the IR-100 Award was submitted by ORNL. That Award was won by Kercel and Bill Dress, in 1997. Some of the illustrations from that submittal are used here. The company Geo-Vox Security, Inc. of Houston TX was licensed by ORNL to develop a commercial version of this technology.

Stephen W. Kercel

ABSTRACT: ORNL has developed an algorithm that enables a device built at the Y-12 Plant to detect the heartbeat of a person hidden in a car or truck. The heartbeat detector, designed to detect hidden intruders before they enter secure nuclear facilities, is also being used to spot escaping prisoners and illegal immigrants concealed in vehicles.

The scene is all too familiar to viewers of real-life TV shows, such as *America's Most Wanted.* Law enforcement has mounted a massive and expensive manhunt, and the countryside is in an uproar. In the latest of what has become a tediously long series of increasingly embarrassing jailbreaks, Willie-the-Axe-Murderer is on the loose. How did he get out? Matching his cleverness with a sudden opportunity, Willie hides in a laundry truck and slips through the prison sally port without being seen by the guards.

While Willie's adventures might be the stuff of bad fiction, even stranger are the true stories of attempts to smuggle passengers concealed in vehicles (see Figs. 1 and 2) through highway border crossings.

ORNL97-3859-C

Fig. 1. Using the heartbeat detector, a technician checks the noise signatures coming from a truck to determine if a person is present inside.

ORNL97-3870-C

Fig. 2. Illustration of a person concealed in truck.

First, there are the fairly straightforward ploys, such as removing the material from inside the front or back seat and stuffing the passenger into the available space. However, the less obvious strategies are nothing short of astounding. For example, U.S. agents recently discovered two illegal aliens wrapped around the engine of a Yugo. The concealment methods attest to both the ruthlessness of the smugglers and the desperation of their illegal but determined customers.

Although it may be debatable just how much the well-being of the nation is threatened by these unusually plucky immigrants, there is another problem that is undeniably quite serious: the risk of terrorist penetration of sensitive areas, such as nuclear facilities. One problem with terrorists is that they are driven by so many different yet powerful motivations. There are militant environmental radicals intent on saving their sacred Mother Earth from the assorted depredations of humankind, religious fanatics seeking to kill infidels, international gangsters in a straightforward search for military-style hardware, and ultra nationalists avenging the wounded honor of their particular tribe. The one attribute that all these people have in common is the belief that if they could get into a nuclear facility, they could really make a statement on behalf of their cause.

Is there anything that security technology can really do about these perils? Can new technology keep prisoners in and illegal immigrants and terrorists out?

Y-12 Engineering and ORNL Algorithm Solve Problem

The Enclosed Space Detection System, known more commonly as the heartbeat detector, provides one solution to these problems. It detects the heartbeat of anyone who might be hiding in a truck, such as the woman shown hiding in Figure 3.

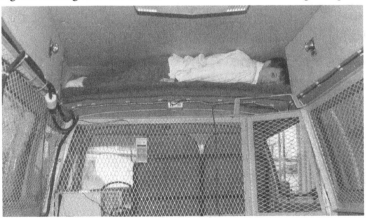

ORNL6811-96

Fig. 3: woman shown hiding in a truck. Almost any space in the cargo compartment of a truck may be used to hide a person. However, the heartbeat of a hiding person is detected by our computer algorithm.

Several years ago, a team of engineers at the Oak Ridge Y-12 Plant, a nuclear facility that stores large quantities of highly enriched uranium, began developing the heartbeat detector as part of the Department of Energy's "Portal of the Future" project. The goal of this project is to create a system that uses a variety of sophisticated devices and methods for rapid inspection of trucks passing through vehicle portals at key facilities. The original purpose of the heartbeat detector was to prevent an intruder hidden in a truck from sneaking through the Portal of the Future, say, to steal nuclear material for weapons or to hold people hostage.

By mid-summer 1995, the heartbeat detector was reliable enough that it was demonstrated at security technology shows all over the country. It astounded audiences wherever it was demonstrated. Particularly surprising was the fact that the heartbeat of a person concealed in a vehicle could actually be detected at the exterior of the vehicle. The detector almost never missed a beat, despite a wide variety of attempts to fool it.

Several years ago, when the heartbeat detector was being developed at the Y-12 Plant, researchers in ORNL's Instrumentation and Controls Division were asked to solve a problem mathematically to ensure that the device would accurately detect human heartbeats amid various background signals. We developed the Fast Continuous Wavelet Transform Algorithm, which is a key to the success of the heartbeat detector.

The heartbeat signal is captured at the exterior of the vehicle by either a geophone (a device commonly used to detect small disturbances in the earth) or a microwave sensor. The sensor signal, which includes truck vibrations from air currents and natural resonances, is fed into the wavelet algorithm, producing the kind of output shown in Fig. 4.

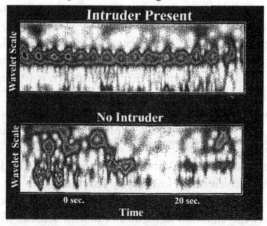

Fig. 4. Wavelet transform of heartbeat signature in noisy geophone output. The top image shows the heartbeat of a concealed person in the vehicle. The bottom image shows no concealed persons in the vehicle.

If a heartbeat is in the signal, it will be matched (and detected) by the heartbeat wavelet programmed into the algorithm. In Fig. 4, the heartbeat stands out of the background noise as a series of easily recognized blobs.

Major Advance in Security Technology

The wavelet-based heartbeat detector has become recognized as a major advance in security technology. It was independently tested at the Thunder Mountain Evaluation Center at Fort Huachuca, Arizona, where it was shown to be more than 99% reliable in detecting occupants hidden in vehicles. In 1996, the Oak Ridge technology was licensed to a private company for development into a commercial product for government and corporate security operations. In 1997, the heartbeat detector received *R&D Magazine's* R&D 100 Award, as one of the 100 most significant technological product developments of the year

The heartbeat detector is the first practical electronic alternative to human or canine inspection of vehicles for hidden passengers; in fact, it is dramatically superior to both inspection methods. An effective human inspection can take hours, possibly requiring the dismantling of an entire vehicle, which is not feasible in most practical settings. By contrast, the heartbeat detector can complete its check of an entire vehicle in less than 2 minutes. The cost of the heartbeat detector is about 3% the cost of canine security, and it is significantly more reliable than the 80% rate of detection that can be expected from a first-class canine security operation.

The system is a dramatic demonstration of how security technology can inexpensively raise the level of public safety. It is already in use at several prisons, and prison officials expect it to reduce the number of jailbreaks. In fact, it prevents many incidents from even being attempted because it decreases the odds of success. In future applications it will be used to discourage terrorist acts and smuggling incidents. One other safety feature is especially important to the end user of the system. Because the heartbeat detector does not require the agent to make physical contact with the intruder, it is safer than human or canine searches, both for the agent and for the intruder.

We envision that this technology will be applied to many other perplexing problems in security technology. For example, car thieves often hide their stolen vehicles in shipping containers and then ship them overseas. It is not practical for customs agents to open every container and inspect it in the detail needed to consistently detect stolen cars. However, by hammering on the container, and then using the heartbeat detector technology to analyze the response, it may be possible to detect the vibration signatures of hidden vehicles without opening the container. Of course, for this application a new wavelet algorithm would be required, and we at ORNL are ready and able to fulfill this need.

Ray Adams

BIOGRAPHICAL SKETCH

ORNL3386-95

Fig. 5. Steve Kercel also received an R&D 100 Award in 1995 for his development with others of the magnetic spectral receiver (shown here), which provides low-cost, highly accurate magnetic field monitoring in facilities where sensitive instruments are used. He and Bill Dress developed the algorithm for the heartbeat detector, which received an R&D 100 award in 1997. Photograph by Tom Cerniglio.

Steve Kercel has been a development engineer in ORNL's Instrumentation and Controls Division since joining the Laboratory in 1990. His research interest is the application of advanced signal processing to security technology, transportation, and power systems. In 1995 he won (jointly with Bill Dress, Mike Moore, and Bob Rochelle) an R&D 100 Award for the application of the wavelet transform to the measurement of broadband radiated magnetic effects on nuclear power plant control rooms. He has a PhD. degree in electrical engineering from the University of Tennessee at Knoxville (UTK), where he is an adjunct faculty member. At UTK he teaches a graduate course on the engineering applications of wavelets. Before joining ORNL, he was a systems analyst in the Office of Inspector General at the Tennessee Valley Authority. He is a registered professional engineer and a senior member of the Institute of Electrical and Electronics Engineers.

285

Chapter 23

Automated Spatial Signature Analysis: Rapidly Tracing Semiconductor Defects To Manufacturing Problems

Kenneth W. Tobin, Shaun S. Gleason, and Thomas P. Karnowski

ABSTRACT: ORNL and SEMATECH have developed an automated spatial signature analysis algorithm for the semiconductor industry that identifies and analyzes distribution characteristics of defects on silicon wafers. It then groups these defects into signatures that can be linked to equipment sources of contamination and fabrication process faults that need to be corrected to improve product yield. The award-winning software tool has been licensed to 14 semiconductor manufacturers and equipment suppliers. It is expected to improve the U.S. semiconductor industry's productivity and cost effectiveness in producing electronic components for computers.

Editor's Note: The illustrations in the original paper benefit from the use of color. If the gray scale nature of the reproductions in this book causes difficulty in your interpretation, Ken Tobin is available to clarify, by calling 865-574-0355

Moor's Law

According to Gordon Moore's law, the density of integrated circuits on a semiconductor chip doubles every two years and the computing speed resulting from jamming more circuits on each chip doubles every 18 months. By compressing the distances that electrons must travel, the semiconductor industry is manufacturing chips that make more calculations faster while using less power. These smaller, faster chips are in great demand by computer makers whose goal is to meet growing business and consumer demands for faster machines. The authors are shown below in Fig 1, as they view the results of this important work.

ORNL2245-98

Fig. 1. Ken Tobin (left), Shaun Gleason, and Tom Karnowski view a display of results from ORNL's spatial signature analysis algorithm.

Just as chip speed increases, so does the amount of data on the chip manufacturing process because the process has become increasingly complex. The process involves the production of integrated-circuit microdevices on 200-millimeter (8-inch) wafers of silicon, which are thin black circular disks dotted with tiny luminescent squares called dies. Each die is an integrated circuit, or a tiny slice of material on which is etched or imprinted a trace that carries current and a complex of electronic components and their interconnections. These dies are removed from the wafer at the end of the manufacturing process and are packaged to make a microprocessor—an integrated circuit that contains the entire central processing unit of a computer on a single chip ranging in size from a baby's fingernail to a quarter.

To produce these dies, 300 to 400 process steps are involved. Each die is formed by deposition of about a dozen layers, including metal films, insulating material, and semiconducting material with electrical contacts to make transistors for uses such as switches. With so many steps and materials involved, defective dies can be formed. For example, a drop of water or a flake of dirt, metal, or polymer from a previous step can end up on the wrong layer (one source of such contamination may be a leak in the chemical vapor deposition vacuum line that shows up as a "spray deposition pattern" on numerous wafer dies). A layer may be missing a circuit pattern or it may be imprinted with an extra pattern. An abnormal ring pattern indicates that too much material was etched away or not enough was deposited. A series of dies might have the same defect stemming from a scratch on the wafer resulting from improper mechanical handling by a robot (see Fig. 2).

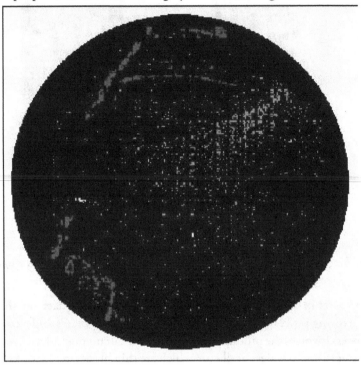

Fig 2. A scratch-and-spray deposition pattern is shown on this wafer map.

The semiconductor industry is concerned with yield, a measure of productivity. Yield is based on the number of wafers processed per hour and the percentage of good dies produced on these wafers. A yield of 70 to 90% good dies is considered desirable. To track the yield of the fabrication process and equipment, the wafers are imaged using in-line laser scanning devices and optical microscopes linked to a computer. Engineers then view these low-resolution images, or wafer maps. If too high a failure rate is observed (based on statistical analysis), selected wafers are sent to the failure analysis lab for off-line defect review. This procedure, in which yield engineers obtain and examine high-resolution optical microscope and scanning electron microscope images of hundreds of wafers each day, is time consuming and expensive. But sorting through all these data has been the only way to determine the source of the defects—usually a process fault or malfunctioning equipment causing contamination, scratches, or stains. Yield engineers try to identify the root cause of the defects by recognizing a known pattern, or "spatial signature"—a unique distribution of wafer defects originating from a single manufacturing problem. Clearly, a data reduction method that quickly extracts only meaningful information relating defects to manufacturing malfunctions is needed. Such a method will help semiconductor manufacturers reach an important goal—"ramping up" the manufacturing process more quickly to achieve the desired yield.

ORNL-SEMATECH

Software Tool

To help semiconductor manufacturers cut costs and increase efficiency and productivity, ORNL and SEMATECH (a research organization of the semiconductor industry) have developed an automated spatial signature analysis (SSA) algorithm for the semiconductor industry that identifies and analyzes defect distribution characteristics on silicon wafers. This artificial intelligence method then groups these defects into spatial signatures that can be linked to manufacturing faults that need to be corrected to improve the quality of product yield. It is expected that SSA technology will improve the U.S. semiconductor industry's productivity and cost effectiveness in producing defect-free dies for computer chips. The software tool has been licensed to 14 semiconductor manufacturers and equipment suppliers.

SSA was developed under a cooperative research and development agreement (CRADA) between the Image Science and Machine Vision Group in ORNL's Instrumentation and Controls Division and the Defect Reduction Technology Group at SEMATECH in Austin, Texas. In 1987 SEMATECH began as a partnership between the government and the semiconductor industry, with 50% funding from member companies and 50% funding from the Defense Advanced Research Projects Agency (DARPA). Since 1996 the organization has been operating without federal funding, working with member companies and suppliers to develop the equipment and material needed to advance existing technologies and to increase efficiency and cost effectiveness. The goal has been to make incremental changes to maintain the semiconductor industry's historical productivity curve that led to smaller, faster, and more widely used computer chips. To keep up this trend and to help its member companies remain competitive, SEMATECH sponsors research to determine which technologies will be needed in the 21st century to make even smaller, faster computer chips.

SEMATECH played a key role in producing the latest edition of *The National Technology Roadmap for Semiconductors*, which spells out the vision of the semiconductor industry. The Semiconductor Industry Association builds consensus on the future technologies that will be required to maintain the past rate of advances over the next 15 years. Past technology development has been rapid, enabling the manufacture of electronic products that offer higher performance at a lower cost each year. Today the industry can pack integrated circuits with critical features as small as 250 nanometers (nm) each on a chip. The industry's goal is 50 nm by 2012. At a critical feature size as small as 100 nm, the industry faces technical challenges that threaten its normal pace of development, so SEMATECH supports research to determine which technologies will be needed to keep up the pace.

According to the *Roadmap*, the number of transistors (integrated circuits) in a microprocessor chip fabricated in 1997 is 11 million, with critical device dimensions of 250 nanometers (nm). By 1999, the number will be 21 million and the size will be down to 180 nm; in 2006, it is projected to be 200 million, with critical device dimensions of 100 nm; by 2012, 15 years from now, the number of transistors in a microprocessor is projected to be 1.4 billion, with critical device dimensions of 50 nm.

During that 15-year span, the number of processing steps will rise from 350 to 600, and the complexity in finding chip defects will increase from one part in 3.8 billion to one part in 640 billion. Clearly, the semiconductor industry will need an automated method like SSA to classify imaged signatures and quickly identify the manufacturing problems causing them in order to reduce the production of defective dies.

According to the *Roadmap*, as semiconductor manufacturing becomes more complex, it will be a challenge to meet the continuing need of manufacturers to "ramp up" more quickly to a "mature yield" of more than 70% defect-free dies. Decreasing this ramp-up time requires a decrease in the time needed to recognize trends in defect production and to quickly trace these "signatures" to specific manufacturing problems. Thus, the *Roadmap* calls for "automated data reduction algorithms (and software tools) to source defects from multiple data sources . . . to reduce defect sourcing time."

There is already a great need for automated data reduction algorithms. For that reason, ORNL has licensed the SSA technology to 14 companies, many of which are SEMATECH member companies. The licensees are eight semiconductor manufacturers—Advanced Micro Devices, IBM, Intel, Lucent Technologies, Motorola, National Semiconductor, Rockwell, and Texas Instruments—and six semiconductor equipment suppliers—Applied Materials, Defect and Yield Management, Inspex, KLA-Tencor, Knights Technology, Inc., and ADE, Inc.

One of the companies commercializing SSA technology is Knights Technology (Sunnyvale, California) under the trade name SPaR™, for spatial pattern recognition software (see Fig. 3).

ORNL2246-98

Fig. 3. Regina Ferrell uses Knight Technology's commercial version of the ORNL-developed SSA tool called SPaR^{TM}.

Designed to help engineers diagnose the root causes of wafer processing defects, this automated method detects and classifies wafer anomalies based on wafer-map defect distributions. The software uses computer vision algorithms and extraction techniques for pattern analysis to identify and classify defect patterns. The defect wafer maps are then transformed into a summary of process events that can help engineers predict and prevent yield loss. By reducing the cost and time needed to identify the root causes of a manufacturing defect, the software therefore minimizes the amount of defective product made before corrective actions are taken in the manufacturing process.

Automated Defect

Data Analysis

Image-based defect detection workstations produce and store tremendous amounts of data as wafer maps. It is costly and time consuming for yield engineers to manually inspect and evaluate these wafer maps to identify causes of the defects. Now, with SSA technology, this process of analyzing both low-resolution and high-resolution images is automated. The software can be used to import wafer-map data generated by standard inspection tools. The SSA algorithm automatically finds patterns, or unique signatures, on a wafer—or even stacks of wafers—from one lot or across lots. By comparing the new signature with known signatures in a data library, the algorithm groups the new defect pattern into an existing classification that is labeled (unless the signature has never before been seen by SSA, in which case it is labeled "unknown"). Each label associates the class of defects with a known manufacturing process problem. This information allows yield engineers to learn more quickly what must be done to modify the semiconductor manufacturing process to get the desired yield.

The SSA artificial intelligence method must be "trained" to recognize new patterns by comparing them with labeled signatures stored in the SSA data library in the computer's memory. If there is a close match between the new signature and one class of stored signatures, the new signature will be labeled as a member of this class and will be an indicator of a specific manufacturing problem. To train our system, we used fuzzy logic in the form of our pair-wise fuzzy k-nearest neighbor (kNN) algorithm for classifying objects segmented from images.

In the spring of 1997, we installed an SSA C++ software system at three U.S. semiconductor manufacturing sites— Intel, Texas Instruments, and Digital Equipment Corporation. A library was then generated representing each company's unique manufacturing problems. These libraries contain knowledge based on yield engineers' observations of correlations between wafer defect patterns and specific manufacturing problems.

We used these data libraries to evaluate the usefulness of the SSA method by analyzing 1827 signatures on 747 wafers in 198 lots at the three sites. The validation exercise was a five-month test of the maturity of the SSA algorithm and library in three different manufacturing environments on three separate products: application-specific integrated circuit (ASIC) chips, dynamic random access memory (DRAM) chips, and static random access memory (SRAM) chips. The data from this exercise indicated that the SSA technology successfully detects and identifies the key signatures most likely to provide meaningful insight into the manufacturing process and to lead to rapid process characterization and correction.

At the beginning and throughout most of the wafer manufacturing process, in-line optical inspection is used to monitor the process. But at the end of the process, the wafers are subjected to electrical testing to ensure that all the semiconducting dies work well electrically. Electrical test data can be displayed in patterns called bin maps (see Fig. 4).

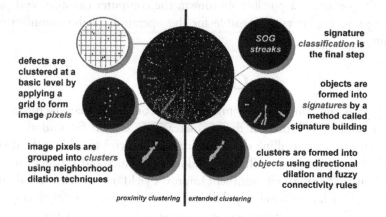

Fig 4a This schematic shows the steps required to ensure a desired yield of defect-free semiconductor wafers

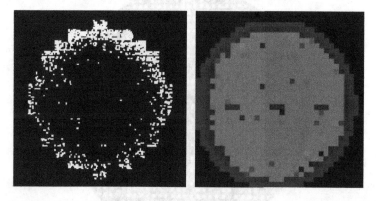

Fig 4b Wafer maps based on optical (left) and electrical testing images (right). The white dies are defect-free but the colored dies have defects (each color represents a particular type of defect).

Deviations from the normal patterns indicate which dies failed the electrical tests. ORNL and SEMATECH collaborators found that SSA technology can be used to process bin map files and see patterns in the electrical test data that indicate which manufacturing process or piece of equipment is faulty. We showed that the SSA algorithm works with electrical test data from Texas Instruments, IBM, and Advanced Micro Devices.

How does the SSA software tool alert engineers that a specific manufacturing problem warrants immediate attention, even a possible shutdown and retooling of the manufacturing process? The computer normally sends the engineers an electronic mail message every 6 to 12 hours to give them information on the process; however, if a defect pattern is severe enough to warrant a possible shutdown, the computer can also send pager messages to engineers responsible for the operation of the manufacturing facility.

Conclusion

Spatial signature analysis provides semiconductor manufacturers with a new level of information that is increasingly critical for understanding, correcting, and controlling the manufacturing process to achieve the desired yield of computer chips. The integration of product and process data into an automated analysis environment improves yield management capabilities. Results of such automated capabilities include improved wafer throughput, rapid root-cause determination, real-time yield analysis, and automatic tool control (see Fig. 5).

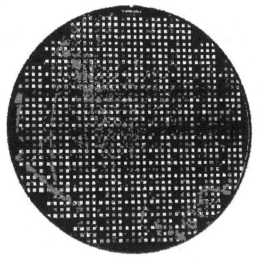

Fig 5 Image of wafer, with defect signatures imposed by
Ken Tobin digitally as design elements

Recent research efforts to validate SSA technology have shown that it is valuable for meeting future yield targets through the reduction of wafer defects generated by a faulty process or malfunctioning equipment. Even more recent research shows that SSA technology is also useful for analyzing electrical test data to gain more insight into the manufacturing process. Our algorithm, which automates the interpretation of wafer defect data, solves the problem of converting massive amounts of process data into useful process control information that will have a positive impact on time to market, product ramp-up time, and wafer fabrication revenue.

Other industries that may benefit from this knowledge-based tool include manufacturers of flat panel displays and optical and magnetic disks. ORNL has an ongoing effort to apply SSA technology to continuously manufactured textile web products. But, for now, SSA technology is best known as a major success story because it resulted from a partnership between a government research laboratory and the semiconductor industry. As an indication of this success, in 1998 our technology received three awards: a Technical Achievement Award from Lockheed Martin Energy Research Corporation, a Marketed Technology Award from the American Museum of Science and Energy, and a Department of Energy Federal Laboratory Consortium Award for Excellence in Technology Transfer.

Chapter 24

Particle-Physics Electronics

Charles L. Britton and Alan L. Wintenberg

ABSTRACT: ORNL has helped develop detector electronics for particle collision experiments at the European Laboratory for Particle Physics (CERN) and at the PHENIX detector at Brookhaven National Laboratory's Relativistic Heavy Ion Collider.

From the very beginning of particle physics research, experimenters have sought larger detectors and better readout instruments to record data from detectors. The very first particle accelerators were simply cathode-ray tubes (Crookes tubes) built in the late 1800s. These crude devices led to the discovery of X rays more than 100 years ago, marking the beginning of a steady increase in the size and complexity of systems used to study the structure and nature of fundamental particles of the atom (e.g., quarks in protons and neutrons) and to determine if theoretically predicted particles (e.g., Higgs boson) actually exist.

In particle physics experiments, atomic and subatomic particles are intentionally speeded up and collided in a controlled manner in specially designed accelerators and cyclotrons called colliders. From an engineering viewpoint, a collider detector is simply a giant image processor that reconstructs the invisible bits and pieces that result from intentional particle collisions. To glean the desired information from these events, experimenters must put many thousands of detector "eyes" around the region of particle interaction. The particles interact with special material in a detector, such as a gas or crystal. The result is an electrical signal proportional to either the energy deposited in the detector by the particle or to the instant of time when the particle passed through the detector. A detector element and associated electronics, called a channel, are then used to convert each electrical signal to digital data for the computer.

The large size of the detectors and the large number of electrical signals that must be processed require many channels of readout electronics. The only way to put tens of thousands of channels into a limited space is to employ integrated-circuit technology, reducing electronics that once occupied tens of cubic centimeters of space to a few cubic millimeters. Thanks to this technology, the PHENIX collider detector at Brookhaven National Laboratory (BNL) has more than 400,000 channels of electronics.

Our group in ORNL's Instrumentation and Controls Division has recently helped to develop detector electronics for particle collision experiments at the European Laboratory for Particle Physics (CERN) near Geneva, Switzerland, and at the PHENIX detector at BNL's Relativistic Heavy Ion Collider (RHIC), as shown in Figs. 1.and 2

A photograph of the PHENIX detector, taken from
http://rsgi01.rhic.bnl.gov/~phenix/GIF/96mag_2.gif

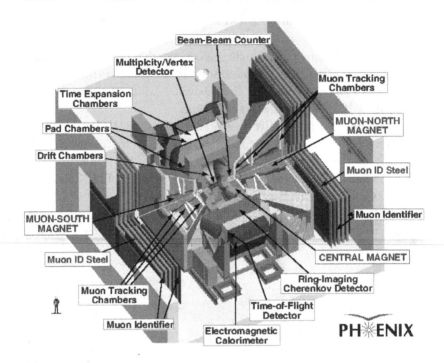

Fig. 2. PHENIX Detector internals. ORNL researchers will perform quark-gluon plasma experiments at the PHENIX detector (shown here) at the Relativistic Heavy Ion Collider (RHIC) at Brookhaven National Laboratory.

CERN Experiment

In 1992 Glenn Young of ORNL's Physics Division came to us and asked our help in writing a proposal to build a 10,000-channel electromagnetic calorimeter for the international WA98 experiment at CERN. A calorimeter is a device that measures the energy of every new particle produced as the accelerated beam strikes a fixed target. One purpose of the WA98 experiment was to try to verify the theory that, within the first 10 seconds of the Big Bang, quarks existed in the free state (in a quark-gluon plasma) before forming the protons and neutrons that became the atomic nuclei of our universe.

We had worked with Young in an earlier project in which we built prototype electronics for a bismuth germinate (BGO) crystal calorimeter. However, the calorimeter needed for the WA98 experiment was to use less expensive lead-glass crystals. Young and his European collaborators already had about 2000 channels instrumented using commercial equipment, but they could not afford to add channels at the prevailing cost of about $300 per channel. They also wanted additional capabilities that commercial electronics lacked. Intrigued, we told Young of our interest in doing the electronics for the lead-glass calorimeter.

During the same time, we were working on integrated electronics for several planned experiments at the Superconducting Super Collider (SSC), which was proposed but never built, and the RHIC, which is being completed. For example, we were developing integrated-circuit (IC) prototypes for one planned RHIC experiment headed by Young. It was recognized that the only way to keep costs down on these experiments requiring thousands of channels of electronics was to develop custom ICs that could each do all or much of the signal processing required by several channels. It appeared to us that this same approach would be needed for the WA98 experiment at CERN (see Fig. 3).

ORNL9147-94

Fig. 3. Chuck Britton (right) shows an integrated-circuit chip that he, Alan Wintenberg (center), and Lloyd Clonts designed at ORNL for use in particle-physics detectors.

The main purpose of the WA98 calorimeter was to measure the energy of particles produced when the beam from the accelerator collided with a fixed target. The interaction of the new particles with the lead-glass crystal produces light of an intensity that is directly proportional to the energy of the incident particle. Each lead-glass crystal has its own photomultiplier tube that converts each light signal to an electrical charge. Because the charge is proportional to the light intensity (which is proportional to the particle energy), measuring charge enables calculation of the energy of the particle. The existing electronics integrate the charge delivered from a photomultiplier and convert it to a digital code that is reported to the data acquisition system. The charge resulting from an event arrives on a submicrosecond time scale. Although the events occur a millisecond or so apart on the average, it is very unlikely that the energy measured is the result of more than one event.

ORNL127938-95

Fig. 4. Glenn Young right) and Alan Wintenberg check a test fixture for testing ORNL-designed integrated-circuit electronics for a silicon detector being developed for the PHENIX quark-gluon plasma experiment at RHIC.

Unfortunately, most of the events observed by such an experiment are not very interesting, so "trigger" logic is used to indicate which events are worthy of a particle physicist's attention. The trigger logic takes the output of several detectors and examines it for, say, total energy deposited or perhaps a pattern. If the result is interesting, then the experiment's data acquisition system must swing into action.

With the commercial charge-integrating, analog-to-digital converter used for the 2000-channel lead-glass calorimeter, taking data was difficult because the electronics had to be triggered before the charge started arriving. Because the trigger detectors are slower than the lead-glass detector, and because the trigger logic requires some time to work, the only option was to delay the arrival of the photomultiplier signal by passing it through a 100-meter delay cable. For the 10,000 channels planned, a million meters were needed. Because this option would considerably boost the cost of the experiment and require a large storage volume, we decided a new approach was needed.

While discussing other ways of integrating the charge, we hit upon a scheme that would use one of the classic nuclear instrumentation circuits—the charge-sensitive amplifier—and one of the newer ones, the analog memory. (We had been designing analog memories for planned RHIC and SSC applications.) The charge-sensitive amplifier would provide the integration function, and the analog memory would provide the delay function. We would design ICs to implement these functions and build them into electronic modules. We estimated that the cost per channel would be less than $100 (one-third of the cost per channel estimated for the first planned data acquisition system).

ORNL938-94

Fig 5a, Integrated-Circuit delay structures replaced thousands of meters of delay cables.

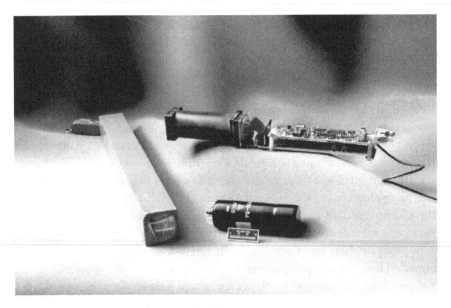

ORNL936-94

Fig. 5b, Examples of detector electronics developed at ORNL

To increase the capabilities of the calorimeter, Young wanted us to add two important features to the electronics—a timing system and a trigger system. The timing system would allow identification of particle type by measuring the particle's time of arrival relative to other channels and other detectors. Because the calorimeter is a considerable distance from the collision point, massless particles traveling at the speed of light, such as photons, arrive sooner than particles with mass, even when they have considerable energy and are traveling at a fair fraction of light speed. The timing system would consist of a constant-fraction-discriminator (CFD) and a time-to-amplitude converter (TAC) in each electronics channel. The trigger system would allow the lead-glass calorimeter to contribute to the trigger logic by summing the signals proportional to the energy deposited in 16-channel sub-arrays of the detector and deciding if any were greater than a programmable threshold. Thus, a deposit of sufficient energy in a small area of the calorimeter could trigger the entire experiment's data acquisition system to capture an event. The project was funded late in 1992, so we started designing the integrated circuits.

Because the whole idea was new, Young urged us to build the prototype chips into a prototype system that could actually be used to take data from a prototype calorimeter exposed to particles from a test beam. We built two 24-channel prototypes, each using 28 custom ICs (four different types), a box full of off-the-shelf ICs, and three large circuit boards. Results from the

October 1993 tests were encouraging technically, but we still wondered how we could finish the design, build all the electronics, and get it installed in time for the October 1994 experiment. When asked if it could be done, Chuck Britton said that the probability of success would be equivalent to driving through 20 traffic lights in a row without encountering a single red light.

By the summer of 1994 we had increased the density of the electronics considerably. The 24-channel prototype had turned into a 144-channel unit requiring 36 custom ICs of only two different types, a large handful of off-the-shelf ICs, and one very large circuit board. As our circuit boards were being produced, the deadline loomed near, so Young's collaborators began inquiring daily about our progress. As soon as a few boards were checked and calibrated, they were shipped to CERN where they were installed with the help of part of the ORNL team. The rest of the team stayed at ORNL to process more boards. In total, 72 boards were installed. The WA98 collaborators were amazed when the calorimeter worked as promised, because their expectations for laboratory- or university-built electronics were much lower than for commercial equipment. We were very pleased to see our IC creations used in large quantity.

PHENIX

The PHENIX detector is a large multi-component detector at BNL's RHIC. PHENIX, which will begin taking physics data by the end of the decade, has more than 400,000 channels of electronics, most of which are implemented as custom-integrated circuits. Young is currently co-spokesman for an experiment planned for RHIC in which our Monolithic Systems Development Group in the I&C Division and much of ORNL's Physics Division are involved. This physics experiment is a quark-gluon plasma experiment similar to the one done at CERN but requiring a more energetic ion beam.

Our group is responsible for electronics for the event vertex-finding subsystem, the pads tracking subsystem, the electromagnetic calorimeter subsystem, and the muon tracking/identification subsystems. We have taken the architecture developed for the WA98 experiment a step further by designing the electronics to take data while reading out data already taken. Such a "dead-timeless" system was developed by using an analog memory that could store four times more data than the one used in the WA98 experiment and by designing a control architecture that coordinates readout with data gathering to prevent the two functions from interfering with each other.

The technologies used for these various subsystems are truly leading-edge. Multichip modules, a technology in which the bare ICs are integrated directly onto a miniature circuit board, are being used by the vertex system to save space and reduce needed power (and thus heat). The latest packaging and surface-mount components and techniques are being used throughout the detector subsystems to ensure maximum reliability and performance and to minimize cost.

We have also been involved in other areas of particle physics. We designed radiation-hardened analog memories for the ill-fated SSC project; for this work we received a 1994 Martin Marietta Energy Systems Technical Achievement Award. We are also working with the Naval Research Laboratories to develop electronics for space-based astrophysics. This is, without a doubt, a most exciting time to be involved in physics instrumentation. The requirements for modern particle-physics experiments present challenges much tougher than those of most other areas of electronic instrumentation.

BIOGRAPHICAL SKETCHES

Charles L. Britton, Jr., is a member of the development staff in the Monolithic Systems Development Group of ORNL's Instrumentation and Controls (I&C) Division. He is also an adjunct assistant professor of electrical engineering at the University of Tennessee at Knoxville, where he earned a PhD. degree in electrical engineering. Before coming to ORNL in 1986, he worked for the Hewlett Packard Company, EG&G ORTEC, and North American Phillips. His main interest is in analog and digital integrated circuit design. He has been active in the research and design of analog-digital semiconductor chips and bipolar integrated circuits for physics and astrophysics experiments and has been involved in developing electronics for experiments at CERN near Geneva, Switzerland, and for the Superconducting Super Collider in Dallas, Texas. He is currently the I&C Division's program manager for electronics development for the PHENIX detector at the Relativistic Heavy Ion Collider in Brookhaven, New York. He is also principal investigator for ORNL's internally funded project on design and readout of microcantilever array sensors. He was co-recipient of a 1994 Martin Marietta Energy Systems Technical Achievement Award for his work on radiation-hardened analog memories. He holds four patents and has more than 40 publications in nuclear electronics and signal processing.

Alan Wintenberg, a native of Oak Ridge, Tennessee, is a member of the development staff of ORNL's I&C Division. Since joining the Laboratory staff in 1990, his main responsibilities have been the development of application-specific integrated circuit (ASIC) electronics for robotics and high-energy physics. He holds a PhD. degree in electrical engineering from the University of Tennessee at Knoxville. He has served as project manager for the Whole Arm Obstacle Avoidance (WAOA) and lead-glass calorimeter electronics for the WA98 quark-gluon plasma experiments at CERN. The WAOA project involved designing custom-integrated circuits and incorporating them into a computer-controlled multisensor system that can collect proximity data in real time to protect a robot arm from collisions. The lead glass project included the development of two complex ASICs which were then incorporated into a system of readout electronics for a 10,000-channel lead glass particle detector. He is currently involved in designing ASICs for PHENIX and for wireless and optical sensing applications. He is an adjunct assistant professor of electrical engineering at the University of Tennessee at Knoxville.

Chapter 25

Conclusion of this Book

Ray Adams

The authors have attempted to provide a view of the Instrumentation and Controls practices that took place at ORNL from the beginnings of the Laboratory in 1943 to when the I&C Division was disbanded in 2001. The perspective of the I&C Division is the viewpoint from which this history is drawn. As the I&C Division included a large and diverse number of people and projects, much has been left out. Yet, as one of the managers, whose tenure exceded mine remarked; "I had no idea such work [as is described in this book] was done."

Attempts were made to include coverage of the fine innovative, inventive early electronic work that preceded availability of any commercial instruments, but no one could be found willing to write about (or who could remember) that early work. The authors regret the omission of coverage of that work and there are doubtless other omissions that we are not even aware of.

When the I&C Division was dismantled in 2001, we were saddened at the loss of what we perceived to be an integrated whole (including Instrument Technician bargaining-unit people) working under the same management and alongside engineering guidance, to solve ORNL technical problems. This unique arrangement existed nowhere else and provided solutions at lower cost and greater speed than any other arrangement. The high degree of interaction of the various subdivisions of the I&C engineering groups provided a greatly beneficial symbiotic interplay for many projects. Our beliefs in the beneficial nature of the fundamental worth of the I&C Division was partially restored in 2007, when many of the parts of the I&C Division were brought back together in the formation of the Measurement Science and Systems Engineering Division. However, the integrated nature of the whole (including Instrument Technician bargaining-unit people) apparently will be forever lost.

Whether or not ORNL Management fully appreciated the work of the I&C Division, other agencies of the U.S. Government did. The

Work-for-Others program supplied the I&C Division with much work - even during otherwise "lean" times of DOE funding. Some of these programs have been mentioned in previous chapters of this book, but the funding (specifically directed to the I&C Division) during the Carter administration, for "conservation" R&D was significant.

Appendix A - Organization Charts & List of Progress Reports

Ray Adams

List of Instrumentation & Controls Progress Reports – ORNL Reports

When the ORNL Reports first began to be published, sometime in 1950, the reports of Instrument Research and Development were gathered together and published on a quarterly basis from several groups. Contributors to these quarterly reports were groups in the Chemistry, Health Physics, and the Physics Division and the Instrument Department of the Engineering and Maintenance Division. After the I&C Division was formed, in 1953, the reports became Semi-Annual, Annual, and Biennially published until they ceased to be published after 1997.

Quarterly Instrument Research & Development Progress Reports

ORNL	Date	ORNL	Date
714	4/15/50	1160	10/20/51
796	7/15/50	1335	1/20/52
924	10/31/50	1336	4/20/52
1021	1/20/51	1389	7/20/52
1056	4/20/51	1436	10/20/52
1159	7/20/51	1492	1/20/53

Semiannual Instrumentation & Controls Progress Reports

ORNL	Date	ORNL	Date
1694	7/31/53	1997	7/31/55
1749	1/31/54	2067	1/31/56
1768	7/31/54	2234	7/31/56
1865	1/31/55		

Annual Instrumentation & Controls Progress Reports

ORNL	Date	ORNL	Date
2480	7/1/57	4091	9/1/66
2467	7/1/58	4219	9/1/67
2487	7/1/59	4335	9/1/68
3001	7/1 60	4459	9/1/69
3191	7/1/61	4620	9/1/70
3378	9/1/62	4734	9/1/71
3578	9/1/63	4822	9/1/72
3782	9/1/64	4990	9/1/73
3879	9/1/65	5032	9/1/74

Biennial Instrumentation & Controls Progress Reports

ORNL	Date	ORNL	Date
5196-7	9/1/74 to 9/1/76	6308	7/1/84 to 7/1/86
5482-3	9/1/76 to 9/1 78	6524	7/1/86 to 6/30/88
5758-9	9/1/78 to 9/1 80	6635/V1-2	7/1/88 to 6/30/90
5931/V1-2	9/1/80 to 7/1/82	6729/V1-2	7/1/90 to 6/30/92
6105/V1	7/1/82 to 7/1/84	6817	7/1/92 to 6/30/94

Last Instrumentation & Controls Division Progress Reports

ORNL	Date	Period
6530	12/31/97	7/1/94 to 12/31/97
(Working Together on New Horizons)		
6531	12/31/97	7/1/94 to 12/31/97
(Publications, Presentations, Activities and Awards)		

Organization Charts

1946-47: Organization

The organization chart in this volume for 1946-47 is a composite made from the organization charts that I found in ORNL archives. The I&C Division did not exist until 1953, so I extracted what I could identify of the organizations that were working in the areas subsequently encompassed by the I&C Division.

Refer to The figure AP 1-1 Composite Organization Chart of Early I&C groups at ORNL

It was easy to find the Instrument Engineers and mechanics who were working for Monsanto, in 1946, in Electronic Development and process systems maintenance, for they were all pretty much in the department run by Hart Fisher, called the Instrument Department, and operating under the Plant Manager, R.G. Thumser.

Likewise, it was relatively easy to find the group doing Instruments for Chemical and Physical Measurements, under Cas Borkowski, who later agreed to be the first Director of the I&C Division when it was formed by Alvin Weinberg, in 1953. I had known they were originally in the Chemistry Division and in a small group under Ellison Taylor.

Both the group under Hart Fisher and the group under Cas Borkowski, existed under the original University of Chicago operation, from the beginnings of ORNL. However, I cannot find an organization chart any earlier than 1946, after Monsanto took over the operating contract for the Clinton Laboratories.

The group I have identified as the Controls Section was harder to find and may not be (as I show them in this chart in 1946-47) the nucleus of the Reactor Controls Section of the I&C Division. In fact, finding the roots of E.P. Epler, one of the first heads of the I&C Division Reactor Controls Department, was a bit of a problem. Tracing Epler from his entry into ORNL led to some speculation, at least partly because he was so reticent to claim recognition and partly because he moved around a bit, upon entry to ORNL. I first see Epler on an Organization chart in 1952, when he was head of a Reactor Controls organization, under A.M. Weinberg, who (in addition to being the Research Head of ORNL) then was head of a Research Director's Department after Carbide & Carbon Chemicals Company became the ORNL operating contractor.

Tracing the Reactor Controls effort from its roots in the control of the Graphite Reactor leads to the fact that only until nuclear reactors capable of "fast" periods were projected, did a nuclear reactor need much control or safety apparatus. In this last regard, the operators of the first controlled nuclear reaction under the stands of the Chicago Stadium were very lucky. That pile as well as the Graphite Reactor were really not capable of a "fast" excursion that otherwise might have gotten them into deep trouble.

I have placed the nucleus of what I think may be the group that preceded the Reactor controls group of the I&C Division in what shows up on the Org Chart of that early time in the Power Pile Division. I have done this, even though Henry Newson, who shows up elsewhere in Wigner's R&D organization, was in a different group from the Power Pile Division. Newson, one of the early supervisors of the initial loading of the graphite pile, and who shows up elsewhere in the Physics Div. at that time, wrote a definitive report, that even today is used as a basis for teaching the elements of reactor control.

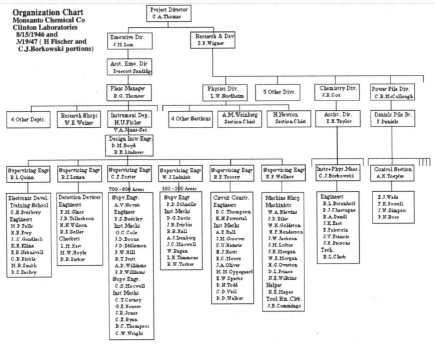

Figure AP 1 - 1: Composit Organization Chart of Early I&C groups at ORNL

1951: Organization

An Instrument Department existed as a part of the Plant and Equipment Division of Union Carbide Nuclear, as early as 1948. I believe that this organization is descended from the organization that supported the chemical re-processing of fuel from the Graphite Reactor in 1946-47. In 1948, D. M. Cardwell was superintendent of the Instrument Department in the Engineering, Maintenance & Construction Division of the Union Carbide Organization. That department handled the maintenance and development of a large share of the Instrumentation of ORNL until the I&C Division was formed. In 1949, C. S. Harrill was brought from the Y-12 plant and was appointed by Dave Cardwell to head the Instrument Department, as Cardwell moved up to division Director of the E&M Division. By 1951, Charlie Harrill had fleshed-out his department to the point that the organization chart looks fairly well populated.

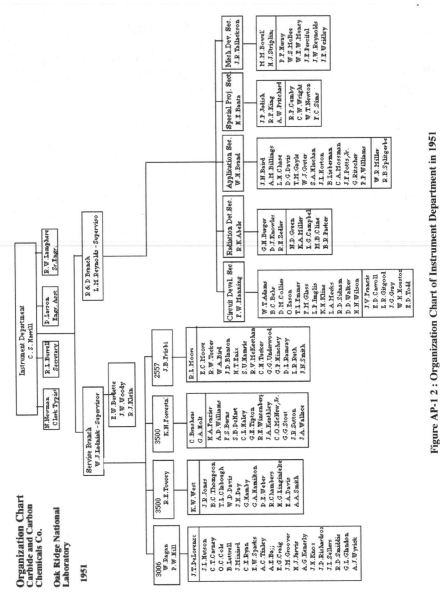

Figure AP-1 2 : Organization Chart of Instrument Department in 1951

1953: Organization

Organization chart of the I&C Division, as the Division was formed in 1953. From 1953 to 2001 there are organization charts presented (about) every 12 years to provide a periodic snapshot of the I&C Division.

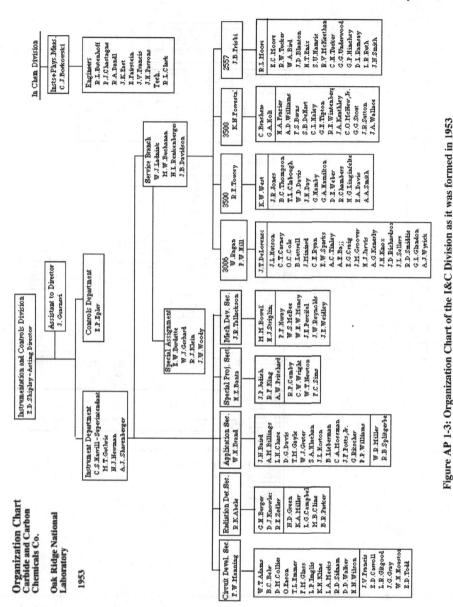

Figure AP 1-3: Organization Chart of the I&C Division as it was formed in 1953

1966: Organization

The I&C Division Organization Chart for 1966 is the first one I found that has many of E.P. Epler's people listed. Ep seemed to have an aversion (over the years) to listing his organization. The organization chart is presented here in two halves.

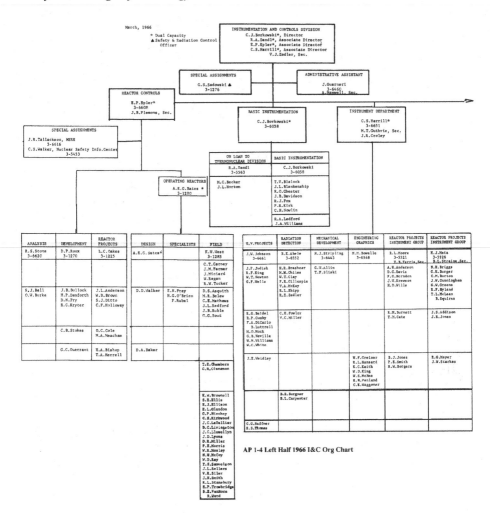

March, 1966

* Dual Capacity
▲ Safety & Radiation Control Officer

INSTRUMENTATION AND CONTROLS DIVISION
C.J.Borkowski*, Director
R.A.Dandl*, Associate Director
E.P.Epler*, Associate Director
C.S.Harrill*, Associate Director
V.J.Zedler, Sec.

SPECIAL ASSIGNMENTS
G.S.Sadowski ▲
3-1276

ADMINISTRATIVE ASSISTANT
J.Guarneri
3-6460
A.Maxwell, Sec.

REACTOR CONTROLS
E.P.Epler*
3-6608
J.B.Flemons, Sec.

BASIC INSTRUMENTATION
C.J.Borkowski*
3-6058

INSTRUMENT DEPARTMENT
C.S.Harrill*
3-6651
M.T.Guthrie, Sec.
J.R.Croley

SPECIAL ASSIGNMENTS
J.R.Tallackson, MSRE
3-6616
C.S.Walker, Nuclear Safety Info.Center
3-3453

ON LOAN TO THERMONUCLEAR DIVISION
R.A.Dandl
3-5563

BASIC INSTRUMENTATION
C.J.Borkowski
3-6058

OPERATING REACTORS
A.E.G.Bates *
3-1280

H.C.Becker
J.L.Horton

T.V.Blalock
J.L.Blankenship
R.O.Chester
J.B.Davidson
R.J.Fox
P.R.Kirk
C.H.Nowlin

R.A.Ledford
J.A.Williams

ANALYSIS	DEVELOPMENT	REACTOR PROJECTS	DESIGN	SPECIALISTS	FIELD		U,V,PROJECTS	RADIATION DETECTION	MECHANICAL DEVELOPMENT	ENGINEERING GRAPHICS	REACTOR PROJECTS INSTRUMENT GROUP	REACTOR PROJECTS INSTRUMENT GROUP
R.S.Stone 3-6620	D.P.Roux 3-1270	L.C.Oakes 3-1225	A.E.G.Bates*		K.W.West 3-1285		J.W.Johnson 3-6461	R.E.Abele 3-6552	H.J.Stripling 3-6443	M.M.Bowello 3-6568	R.L.Moore 3-5521	H.J.Metz 3-5926
					C.T.Carney J.M.Farmer J.Miniard W.Ragan R.W.Tucker		J.P.Judish R.F.King W.T.Newton G.F.Wells	H.R.Brashear H.M.Chiles W.T.Clay F.E.Gillespie V.A.McKay R.L.Shipp R.E.Zedler	C.V.Allin T.F.Sliski	V.N.Ferris,Sec. A.N.Anderson D.C.Davis P.C.Herndon J.W.Krewson H.D.Wills	B.L.Strainc,Sec. N.H.Briggs C.H.Burger C.M.Burton J.W.Cunningham G.W.Greene R.F.Hyland T.L.McLean B.Squires
S.J.Ball O.W.Burke	J.B.Bullock M.P.Danforth D.N.Fry R.C.Kryter	J.L.Anderson W.D.Brown S.J.Ditto C.F.Holloway	D.D.Walker	E.N.Fray H.G.O'Brien F.Rubel	D.S.Asquith M.R.Belew C.E.Mathews J.L.Redford J.B.Ruble N.G.Sout						
	C.B.Stokes	O.C.Cole M.A.Meacham			R.G.Beidel R.P.Cunby T.A.DiCarlo B.Luttrell H.O.Mock G.B.Neville W.H.Williams W.C.White			C.E.Fowler V.C.Miller			R.M.Burnett T.M.Cate	J.S.Addison J.R.Jones
	G.C.Guerrant	H.A.Bishop T.A.Herrell	D.A.Baker				J.E.Weidley			W.F.Greiner R.L.Hansard E.C.Keith W.D.King W.S.McRee R.N.Penland C.E.Waggoner	B.J.Jones P.E.Smith R.W.Borgers	E.G.Meyer J.W.Scarbro
					T.E.Chambers C.M.Cinnamon							
					K.W.Brownell S.H.Ellis H.J.Ellison E.L.Glandon G.P.Hinckey C.E.Kirkwood J.C.LaBellier B.C.Llewellyn J.D.Lyons D.R.Miller P.E.Morris W.R.Mosley M.W.McCoy W.D.Ray T.E.Samuelson J.L.Sellers V.R.Siler J.N.Smith R.L.Stansbury R.P.Trowbridge R.E.VanBorn R.Ward		B.R.Burgner B.L.Carpenter					
							C.G.Ruffner R.S.Thomas					

AP 1-4 Left Half 1966 I&C Org Chart

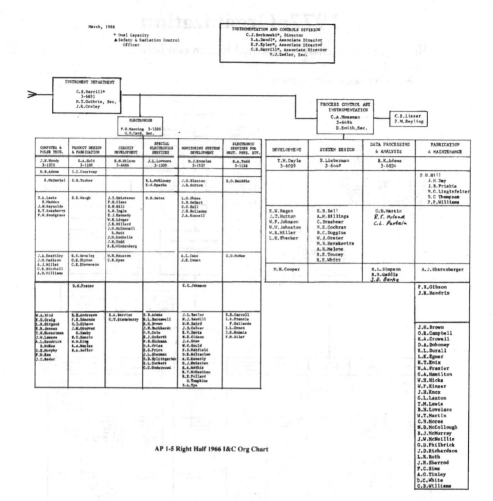

AP 1-5 Right Half 1966 I&C Org Chart

1977: Organization

The organization chart is presented here in two halves.

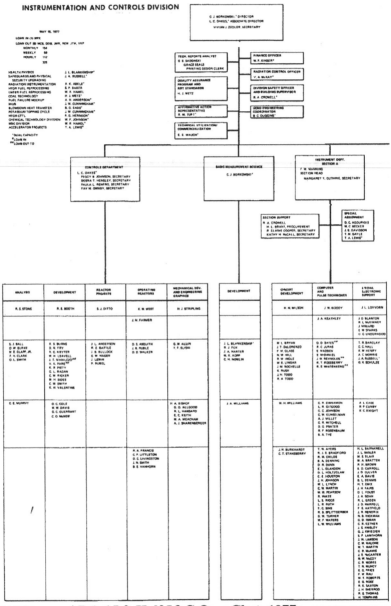

AP 1-6 Lft Half I&C Org. Chrt. 1977

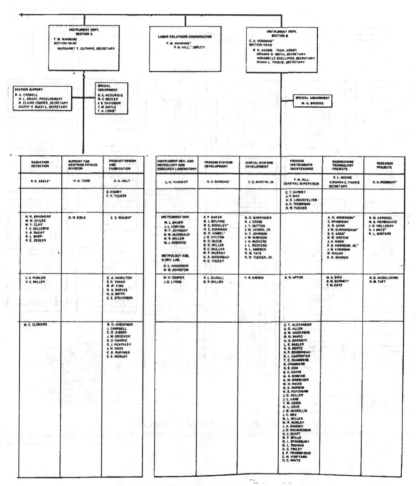

Rght Half 1977 I&C Org. Chrt

1989: Organization

The I&C Division had grown to its largest size, almost 350 persons, by 1989. The organization chart is presented here in two halves.

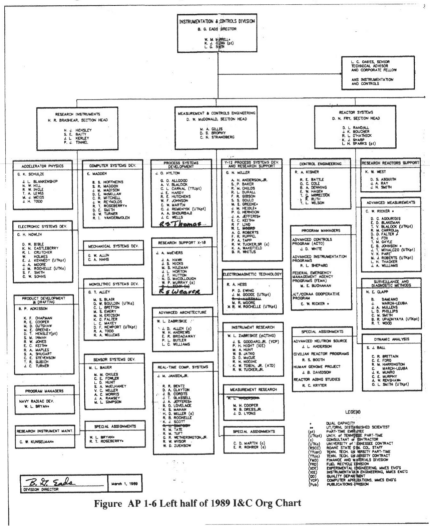

Figure AP 1-6 Left half of 1989 I&C Org Chart

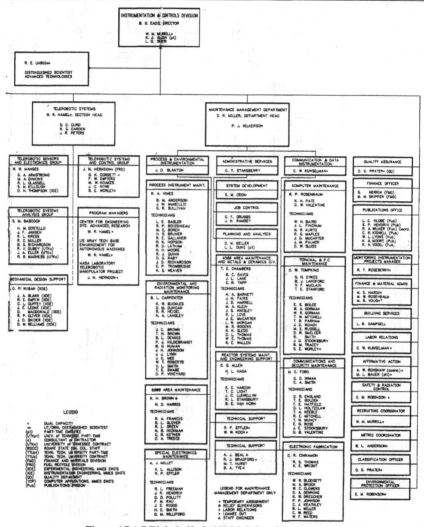

Figure AP 1-7 Right half of 1989 I&C Org Chart

2001: Organization

The I&C Division Organization Chart as the Division was to have been organized prior to the massive re-organization of ORNL in 2001, by the new operating contractor, UT-Battelle. That re-organization eliminated several long-standing ORNL divisions in addition to the I&C Division.

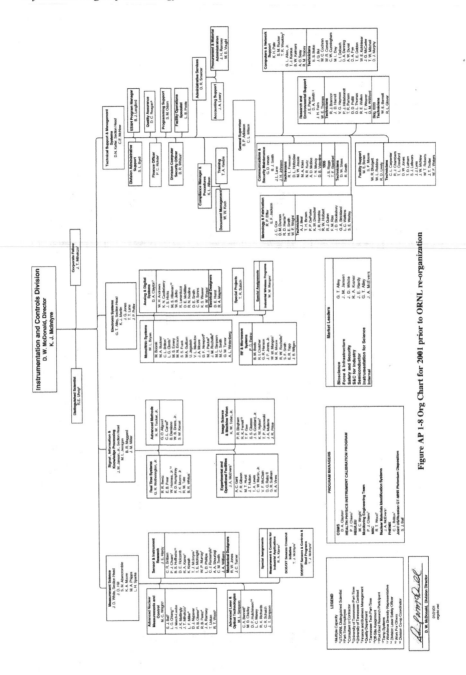

Figure AP 1-8 Org Chart for 2001 prior to ORNL re-organization

Appendix B -I&C Division R&D 100 Winners

Ray Adams

Note: Prior to 1987, the name of the award was "I-R 100 Award". The awards were published in the Industrial Research Magazine. The award has substantial merit, given that it is based upon the 100 most outstanding Research and Development efforts in the whole United States. Of the 139 awards presented to the Oak Ridge R&D community (three organizations), from 1967 to 2002, 21 were given as a result of efforts in the I&C Division. The I&C recipients are as shown below.

2005

This award was based on the work that Don Bible, assisted with while the Instrumentation and Controls Division existed.

Engineering Science and Technology and SensArray Corporation (Fremont, CA)

JOINT w/: SensArray Corporation (Fremont, California)

Carl W. Sohns (ORNL), Robert J. Lauf (Consultant), Don W. Bible (Retired from The Instrumentation and Controls Division), Wayne Renken (SensArray Corporation), Earl Jensen (SensArray Corporation), Brian Paquette (SensArray Corporation), Jeff Parker (SensArray Corporation) and Jim Barnett (SensArray Corporation)

SensArrayÆ INtegrated Wafer

2002

This award was based on the work that Tobin et.al., did while the Instrumentation and Controls Division existed.

Engineering Science and Technology; Joint with: Applied Materials, Inc. ORNL: Kenneth W. Tobin, Thomas P. Karnowski and Regina K. Ferrell; Applied Materials, Inc.: Amos Dor, Barry Wong and Yifah Gavra

DSI TM - AIR: Defect Source Identifier - Automated Image Retriever

2000

Joint w/ Orbital Sciences Corporation, MSP Corporation, Colorado School of Mines, and U.S. Army Soldier and Biological Chemical Compound

Chemical and Analytical Sciences, Instrumentation and Controls, Computational Physics and Engineering, Life Sciences, and Computer Science and Mathematics (and LMES Advanced Technologies)

Wayne H. Griest (Chemical and Analytical Sciences), William H. Andrews (Instrumentation and Controls), Don W. Bible (Instrumentation and Controls), J. Eric Breeding, Michael N. Burnett (Chemical and Analytical Sciences), Kim N. Castleberry (Instrumentation and Controls), Dwight A. Clayton (Instrumentation and Controls), Richard I. Crutcher (Instrumentation and Controls), Kevin J. Hart (Chemical and Analytical Sciences), Mike S. Hileman (Instrumentation and Controls), Ralph H. Ilgner (Chemical and Analytical Sciences), W. Bruce Jatko (Instrumentation and Controls), Roger A. Jenkins (Chemical and Analytical Sciences), Stephen A. Lammert (Chemical and Analytical Sciences), David E. McMillan (Instrumentation and Controls), Randy L. McPherson (Chemical and Analytical Sciences), Roosevelt Merriweather (Chemical and Analytical Sciences), Richard W. Reid (Computational Physics and Engineering), Irene F. Robbins (Computational Physics and Engineering), David E. Smith (Instrumentation and Controls), Robert R. Smith (Chemical and Analytical Sciences), Carl W. Sohns (Instrumentation and Controls), K. Ann Stewart (LMES Advanced Technologies), Cynthia L. Terry (Computational Physics and Engineering), Cyril V. Thompson (Chemical and Analytical Sciences), Arpad A. Vass (Life Sciences), Robert A. Whitaker (Computational Physics and Engineering), Marcus B. Wise (Chemical and Analytical Sciences), Dennis A. Wolf (Computer Science and Mathematics), R. Wes Wysor (Instrumentation and Controls), and Judy C. Zager (Computational Physics and Engineering)

Orbital Sciences Corporation:

Shephard T. Girion, Francis Dompier, William S. Donaldson, Hsienchi William Niu, Gus Norton, Mike Phillips, Gerry Stillman, and Harry Tamme

MSP Corporation:

Darryl L. Roberts, Benjamin Y. H. Liu, Virgil A. Marple, and Francisco J. Romay

Colorado School of Mines:

Kent J. Voorhees, Franco Basile, Michael J. Beverly, Chris Abbas-Hawks, and Allen B. Henderson

US Army Soldier and Biological Chemical Command:

Alexander Hryncewich and David Sickenberger
The Block II Chemical Biological Mass Spectrometer (Block II CBMS)

1999

ORNL Life Sciences, Instrumentation and Controls, Post Doc, Graduate Student

Vo-Dinh, Tuan; Wintenberg, A. L.; Ericson, M. N.; Alarie, J. P.; Isola, Narayan (post doc); Askari, M.(Grad Student); Miller, G. H.

The Multifunctional Biochip

1999

ORNL; Joint w/ American Iron and Steel Institute (Joseph Vehec), Bailey Engineers, Inc. (Richard A. Barcelona), and National Steel Technical Center (Liwei Zhang)

Engineering Technology; Instrumentation and Controls

Allison, Stephen W.; Beshears, David L.; Cates, Michael R.; Childs, R. Mitchell; Manges, W. W.; McIntyre, Timothy J.; and Simpson, Marc

The Galvaneal Temperature Measurement System (GTMS)

1997

ORNL/Lambda Technologies

Instrumentation and Controls; Metals and Ceramics

Bible D; Lauf R;   Lambda Technologies: Fathi Z, hampton M, and Stevens R

Vari-Wave, Microwave Heating Instrument

1997

LMES/ORNL

National Security Programs Office; Protective Services; Instrumentation and Controls; Engineering

Labaj; Bath; Baylor; Carroll; Dress; Fuller; Hickerson; Kercel; McCoig; Pack

Enclosed Space Detection System

1995

ORNL

Instrumentation and Controls

Remenyik CJ - Hylton JO - McKnight TE - Hutchens RE, Gravimetric Gas Flow Calibrator

1995
 ORNL
 Instrumentation and Controls
 Kercel SW - Dress WB - Rochelle RW - Moore MR
 Magnetic Spectral Receiver

1993
 ORNL Engineering Physics and Mathematics - Robotics and Process
Systems
 Pin FG - Killough SM
 Omnidirectional Holonomic Platform (OHP)

1991
 ORNL Instrumentation and Controls/Metals and Ceramics
 Hoffheins BS/Lauf RJ
 Rapid Fuel Analyzer

1988
 ORNL Instrumentation and Controls/A. G. Technical Assoc., Inc.
 Butler P-Allen J (Consultant) OPSNET – A New Parallel Computer
Architecture for Expert Systems

1987
 ORNL Instrumentation and Controls
 Mossman CA-McNeilly DR-Jatko WB-Anderson RL-Miller GN
 Remote Sensor and Cable Identifier

1986
 ORNL/EXT Metals and Ceramics/Instrumentation and Controls/
Carnegie-Mellon U Lauf RJ Hoffheins BS-Emery MS-Siegel MW (Carnegie-
Mellon U., Pittsburgh) (Work funded by Cabot Corp.) Integrated Gas
Analysis and Sensing (IGAS) Chip

1985
 ORNL Instrumentation and Controls
 Davidson JB-Case AL
 EIDEC (Electronic Image Detector for Electrophoresis and
Chromatography)

1985

ORNL Instrumentation and Controls/Analytical Chemistry
Todd RA-Ramsay RS
Pulsed Helium Ionization Detector Electronics System (PHIDELS)

1984

ORNL Instrumentation and Controls/Fuel Recycle
Satterlee PE-Martin HL-Herndon JN
Sargent Industries, Central Research Laboratories Div. (CRL), Model
M-2 (Control System)

1984

ORNL Solid State/Instrumentation and Controls
Mook HA-Schulze GK
Ultrasonically Pulsed Neutron Time-of Flight Spectrometer

1982

ORNL Instrumentation and Controls
Miller GN-Anderson RL-Rogers SC
Reactor Core Cooling Measurement System

1977

ORNL Chemical Technology/Instrumentation and Controls
Scott CD-Mrochek JE-Genung RK-Johnson WF-Bauer ML-Burtis CA-
Lakomy DG
Portable Centrifugal Fast Analyzer

1977

ORNL Instrumentation and Controls/U of Tenn
Borkowski CJ-Blalock TV Johnson Noise Power Thermometer (Industrial
Prototype System, IPS-2)

Endnotes (by chapter):

Introduction

1 Wm. Thompson: Do a "**Google©**" search for "Lord Kelvin - William Thompson"

2 ISA: http://www.isa.org

3 IEEE: http://ieee.org/

4 Early Instrument Department Organization Chart

See Appendix A figure AP1-2

Chapter 1

5 Brooksbank, et al, ORNL TM 12720 Historical and Programmatic Overview of Building 3019

Chapter 2

6 The Atomic Energy Commission (AEC) was established in 1947. Its classification categories were primarily "Q" and "L". It came into being following the Manhattan Project days. During the Manhattan Project days, There were several classifications, The top two were "Secret" and Top Secret" followed by "Confidential"

Chapter 4

7 See Chapter 14 (L. C. Oakes comments) for more details on ORACLE I, the first ORACLE

8 Earle W. Burdette and Rudolph J. Klein - A Decade of ORACLE Experience, paper presented at Pioneer Days: The First Generation (of computers) in Illinois, Margaret K. Butler, Track Chair, ANL Argonne, IL Presented at the 1985 NCC, Pioneer Days, see Walter M. Carlson Papers (CBI 114), Charles Babbage Institute, University of Minnesota, Minneapolis.

9 See: http://cs.wwc.edu/~aabyan/LABS/AR/IAS.html and other similar WWW pages per Google© search on "IAS Computer"

10 C.T. Fike, October 1959 Communications of the ACM, Volume 2 Issue 10

Chapter 9

11 M. J. Roberts and T. G. Kollie, *Rev. Sci. Instru.*, 48, 1197, (1977).

12 Kollie, T.G., Anderson, R. L., Horton, J.L., and Roberts, M.J., . 48, 501, (1977).

13 Measurement of B versus H of Alumel from 25 to 180 °C, J. L. Horton and T. G. Kollie, 48, 4666 (1977); DOI:10.1063/1.323530

Chapter 12

14 Computer Sciences Division

15 Nuclear Division Security and Safeguards.

Index

A

A-1 Amplifier 39
Abele, R. K. 62
Abele, Roland 52, 87, 88, 90, 131, 193
Ackermann, N.J. 211, 225, 237, 240
Adams, Ray iii, 2, 3, 5, 9, 16, 17, 18,
21, 40, 79, 97, 110, 111, 132,
177, 178, 192, 199, 200, 211,
213, 258, 261, 306, 309, 323
Affel, Bob 36, 58, 178, 186
Affel, R. G. 56
Aircraft Reactor Experiment 37, 69
Alexander, C. T. 62
Alexander, Nancy 45, 46
Alexeff, Dr. Igor 110
Allen, C. G. 62, 140
Allen, Charlie 68
Allin, Wilbur 179
Allison, K. L. 154
American Nuclear Society 6, 72, 109,
249
AMFIP 201, 202, 203, 204, 205, 206,
207, 208, 209, 210
Anderson, (Dick) R. L. iii, 17, 111
Anderson, J.L. 224
Anderson, John 67
Anderson, Joy 1
Anderson, R. H. 62
Andrews, Andy 203
ANP Project 55
ANS 6, 7
Arnette, Ruth 45
ASM 167, 168, 169, 170, 173
Asquith, D. S. 57
Atta, Susie 45
AVIDAC 42, 43
Ayers, T. W. 58

B

Babbage, Charles, Institute 329
Babcock, Scott M. 170
Baird, N. 62
Ball, A.E. 32
Ball, S.J. 254, 257
Ball, Syd 197
Banta, Gene 81, 193
Barclay, Tom 179
Barnawell, H. L. 62
Barnawell, Howard 132
Barnett, H. A. 62
Barnett, Howard 132
Basler, J. L. 62
Basler, John 131
Basler, Larry 132
Basler, L. E. 62
Battle, R.E. 257
Bauer, Martin 97
BDHT 122, 123, 124
Beal, A. J. 68, 279
Beall, Samuel 71
Bell, Earl 197, 199
Bell, P.R. 250, 257
Bentzinger, D.L. 279
Bentz, R. R. 62
Bettis, Ed 36, 50, 58, 216
Betz, Nancy 45, 46
Billings, Arville 197
Bird, William 31
Blair, M. S. 62
Blakeman, E.D. 279
Blalock, Vaughan 110
Blankenship, Jim 134
Blodgett, W. R 153
Blosser, T. V. 94
Bodenheimer, Robert 110
Boissineau, R. P. 62
Booth, R. 254
Boren, M. E. 153

Borkowski, (Cas) Casimer 21, 34, 38, 52, 53, 132, 192, 311
Bouldin, Donald 110
Bradford, James 160
Bradford, R. J. E. 58
Brand, Warren 6, 193
Brashear, Hugh 88, 90, 93, 94, 96
Bratten, W. A. 62
Breeding, Eric 273, 324
Breeding, J.E. 279
Britton, Charles L. 261, 296, 305
Brock, M. 279
Brophy, Debbie 160
Brown Recorder 59, 195
Brown, R. H. 62, 63
Brown, R.H. 62, 63
Bruner, Scott 160
Bryan, Bill 74
Bryan, W.L. 279
BSR 69, 138, 215, 217, 221, 253
Buckley, Page 51
Buckner, Mark 88
Buhl, A.R. 240
Bullock, J.B. 224
Burdette, Earle W. 329
Burger, Howard 82, 146
Burgner, B. 32
Burke, O.W. 249
Burkhardt, J. H. 57
Busing, Bill 49, 84
Butler, Margaret K. 329

C

calculators, card-programmed 41
Campbell, Jack 62
Cardwell, Dave 34, 313
Carney, C. T. 62
Carpenter, Ben 53, 132
Carpenter, B. L. 62, 153
Carroll, E. D. 62
Carter, Jimmy 139
Case, A. L. 62
Castleberry, K.N. 279
Catron, B.G. 37

C&CCC 31
CERN 134, 296, 297, 298, 299, 303, 305
CESAR 169, 171
Chambers, R. 62
Chambers, T. E. 62
Chambers, Troy 53, 132
Chase, Larry 194
Chester, Marilyn 47
Childs, R. M. 58
Chiles, Marion 86, 89, 91, 94
Cinnamon, C. R. 58
Cinnamon, Ray 74, 75, 76
Circuits and Systems Society 6
Clabo, Tom 34, 52
Clapp, N.H. 254
Clark, F.H. 254, 257
Clay, W.T. 96
Clinton Engineering Labs 7
Clonts, L. 279
Clowers, Bill 94, 96
Clowers, W. C. 62
Cole, O. C. 51
Cole, Tom 215
Collins, Ed 132
Control Systems Society 6
Cook, J.A. 30
Cook, Kelsey 112
Cooper, Bud 74
Cooper, R.E. 279
Coveyou, Bob 40, 45
Cox, J.C. 279
Cox, R. E. 62
CRL Model M2 Servo Manipulator 166
Crutcher, R.I. 279
Cuddy, Mike 162, 163
Culkowski, Arlene 45
Culler, Floyd 26
Culver, J. D. 62
Cumby, R. P. 62
Cutshaw, M.D. 279

D

Dandl, R.A. 252, 257
Danforth, H.P. 224
Davidson, J. 33
Davis, Brent 53, 132
Davis, Dell 197
Davis, E. 34, 52
Day, Jim 34, 52, 76, 130
DeHart, Sanford 131
DeLorenzo, Joe 81
Denkins, M. A. 62
Denning, Burt 47, 49
Dennis, B. L. 62
Deporter, P.P. 279
Dilworth, B. 38, 54, 107
Dismuke, Nancy 45
Dresher, K.W. 62
Dress, Bill 280, 285
Duggins, Baden 97
Duncan, D.M. 153, 279
Dunn, D. R. 58

E

Eads, B. G. 3, 146
Eddlemon, J. D. 32
Effler, Robert 153, 156, 158, 164
Elza Gate 35
E&M Division 30, 128, 313
Emery, M.S. 279
Emmett, Margaret 46
England, D.R. 279
Enix, H. T. 62
Epler, (Ep) E. P. 15, 192, 211, 213,
    224
Etheridge, Y. H. (Young) 186
Ettingshausen-Nernst effect 123, 124
Evenson, T.G. 32

F

Fair, M. 31
Fairs, J. H. 62
Fairstein, Ed 38, 39, 79
Farris, Virginia 50

Felbeck, George 71
Ferrell, Regina 291
FFTF 228, 236
Fike, C. T. 329
Finchum, Lawrence 158
First Aid 33
Fission Products Pilot Plant 196
Ford, B. 32
Forrestal, K. N. 32
Foust, D. L. 62
Fowler, C. E. 62, 96
Fox, Ted 1
Francis, John 79
Francis, R. A. 57
Frazier, H. 34, 52, 59, 76, 130
Freer, Eva 207
Frisbie, J. B. 29, 61, 132
Fry, D.N. 211, 241, 249, 260
Fry, Dwane 258, 259

G

Gallaher, R E. 153
Garrison, Arlene 112
Gayle, Tom 52, 97, 193, 195
Gehl, A.C. 279
Geiger-Mueller 30
Geo-Vox 280
Gerhard, Bill (William J.) 42, 43
Gibson, C. D. 62
Gifford, Frank 216
Gillespie, Frank 87, 88, 91, 94
Gitgood, L. R. (Ralph) 47, 49, 58, 79,
    131
Glass, F. 38, 39, 54, 79, 81
Gleason, Shaun S. 261, 286
Goan, J. A. 62
Goans, John 131
Gonzales, Ralph 110
Googe, J.M, Dr. 110
Gorman, Bill 158
Goss, G.C. 33, 34, 35
Graphite Pile 24, 29

338